Android Studio
移动开发教程

肖 琨 吴志祥 史兴燕 张 智 编著

电子工业出版社
Publishing House of Electronics Industry
北京·BEIJING

内 容 简 介

本书系统地介绍了在 Android Studio 3.5.x 环境下开发 Android 应用的基础知识和实际应用。全书分 11 章，包括 Android 应用开发概述及开发环境、Android 项目文件结构分析及调试、Android UI 与 Activity 组件、手机基本功能程序设计、服务组件及其应用、广播组件与通知、SQLite 数据库编程、Android 内容提供者组件、Android 近距离通信技术、位置服务与地图应用开发和 Android 网络编程。

本书以介绍 Android 的四大组件为主线，精心组织目录和案例，并在 Android 7.0 手机（或模拟器）上成功运行。此外，每章都精心设计了典型案例以说明其用法，并配有习题及实验。本书配套的教学网站，包括教学大纲、实验大纲、各种软件的下载链接、课件和案例源代码下载链接、在线测试等内容，极大地方便了教与学的实施。

未经许可，不得以任何方式复制或抄袭本书之部分或全部内容。
版权所有，侵权必究。

图书在版编目（CIP）数据

Android Studio 移动开发教程 / 肖琨等编著．—北京：电子工业出版社，2019.1
ISBN 978-7-121-34287-5

Ⅰ．①A… Ⅱ．①肖… Ⅲ．①移动终端－应用程序－程序设计－高等学校－教材 Ⅳ．①TN929.53

中国版本图书馆 CIP 数据核字（2018）第 111032 号

策划编辑：张小乐
责任编辑：王　炜
印　　刷：北京虎彩文化传播有限公司
装　　订：北京虎彩文化传播有限公司
出版发行：电子工业出版社
　　　　　北京市海淀区万寿路 173 信箱　邮编：100036
开　　本：787×1092　1/16　印张：19.5　字数：499 千字
版　　次：2019 年 1 月第 1 版
印　　次：2024 年 12 月第 13 次印刷
定　　价：55.00 元

凡所购买电子工业出版社图书有缺损问题，请向购买书店调换。若书店售缺，请与本社发行部联系，联系及邮购电话：（010）88254888，88258888。
质量投诉请发邮件至 zlts@phei.com.cn，盗版侵权举报请发邮件至 dbqq@phei.com.cn。
本书咨询联系方式：（010）88254462，zhxl@phei.com.cn。

前　言

　　Android 一词的本义是"机器人"，是由 Google 公司于 2007 年 11 月对外发布的一种以 Linux 为基础的开源操作系统，主要用于移动设备。近年来，Android 平台得到了广大手机厂商和移动运营商的广泛支持。Android 智能手机的强大功能和广泛普及，促使各高校纷纷开设 Android 移动平台开发课程。

　　本书系统地介绍了在 Android Studio 3.5.x 环境下开发 Android 应用的基础知识和实际应用。全书分 11 章，包括 Android 应用开发概述及开发环境、Android 项目文件结构分析及调试、Android UI 与 Activity 组件、手机基本功能程序设计、服务组件及其应用、广播组件与通知、SQLite 数据库编程、Android 内容提供者组件、Android 近距离通信技术、位置服务与地图应用开发和 Android 网络编程。

　　本书以介绍 Android 的四大组件为主线，对于章节中的很多知识点，本书都精心设计了典型案例以说明其用法，并配有习题及实验。本书配套的教学网站，包括教学大纲、实验大纲、各种软件的下载链接、课件和案例源代码下载链接、在线测试等内容，极大地方便了教与学的实施。

　　本书写作特色鲜明：一是教材结构合理，对教材目录的设置进行了深思熟虑，多次推敲，在正文中指出了相关章节知识点之间的联系；二是知识点介绍简明，编者精心设计的案例紧扣理论；三是采用大量的截图，可清晰地反映 jar 包、软件包、类（或接口）的层次关系；四是通过综合案例的设计与分析，可让学生综合使用 Android 应用开发的各个知识点；五是开发制作了配套的上机实验网站，方便教与学的实施。

　　本书第 1～5 章由肖琨编写；第 6～11 章由吴志祥、史兴燕和张智编写。吴志祥负责课程教学网站的开发，张智负责课件的制作。

　　本书既可作为高等院校计算机专业和非计算机专业学生学习"Android 移动平台应用开发"等课程的教材，也可作为 Android 初学者的入门参考书。

　　访问本书配套的课程网站 http://www.wustwzx.com/as/index.html，可获取课件、案例源代码等教学资料。

　　由于编者水平有限，书中错漏之处在所难免，在此真诚欢迎读者多提宝贵意见，读者可通过访问编者的教学网站 http://www.wustwzx.com 与编者 QQ 联系，以便再版时更正。

编　者
2018 年 10 月于武汉

目 录

第1章 Android 应用开发概述及开发环境 ... 1
1.1 移动开发与智能手机 ... 1
- 1.1.1 移动开发概述 ... 1
- 1.1.2 Android 智能手机的使用特点 ... 1
- 1.1.3 手机智能操作系统及分类 ... 2

1.2 Android 系统架构 ... 2
- 1.2.1 Linux 内核层 ... 3
- 1.2.2 函数库和 Android 运行时环境层 ... 4
- 1.2.3 应用程序框架层 ... 4
- 1.2.4 应用程序层 ... 5

1.3 Android Studio 开发环境 ... 5
- 1.3.1 Android Studio 概述 ... 5
- 1.3.2 下载和安装 Android Studio 3.1.2 ... 5
- 1.3.3 Android Studio 相关文件夹 ... 6
- 1.3.4 Android SDK 与 Android API ... 6
- 1.3.5 Android Studio 常用组合键 ... 10

1.4 创建和运行 Android 应用 ... 10
- 1.4.1 创建一个 Hello 级 Android Studio 项目 ... 10
- 1.4.2 手机模拟器与 AVD Manager ... 14
- 1.4.3 安装和运行 Android 应用 ... 16

习题 1 ... 17
实验 1 ... 18

第2章 Android 项目结构分析及调试 ... 19
2.1 Android 项目的文件系统结构 ... 19
- 2.1.1 项目配置清单文件夹 manifests ... 19
- 2.1.2 源程序文件夹 Java ... 20
- 2.1.3 资源程序文件夹 res ... 20
- 2.1.4 项目多模块及构建 Gradle Scripts ... 22
- 2.1.5 使用 Project 或 Packages 视图 ... 24

2.2 Android 应用程序的基本组成 ... 25
- 2.2.1 Activity 组件与视图 View ... 25

2.2.2　Service 组件 ………………………………………………………………………… 25
　　2.2.3　BroadcastReceiver 组件 ……………………………………………………………… 25
　　2.2.4　ContentProvider 组件 ………………………………………………………………… 26
　　2.2.5　Application、Context 和 Intent ……………………………………………………… 26
　　2.2.6　Android 应用程序的运行入口 ………………………………………………………… 27
2.3　Android 虚拟机 Dalvik …………………………………………………………………………… 28
2.4　Android Studio 项目调试 ………………………………………………………………………… 29
　　2.4.1　主动调试（Toast 与 Logcat）………………………………………………………… 29
　　2.4.2　动态调试 ……………………………………………………………………………… 31
　　2.4.3　单元测试 ……………………………………………………………………………… 32
习题 2 ……………………………………………………………………………………………………… 35
实验 2 ……………………………………………………………………………………………………… 36

第 3 章　Android UI 与 Activity 组件 …………………………………………………………… 37

3.1　用户界面 UI 设计 ………………………………………………………………………………… 37
　　3.1.1　Android 界面视图类 ………………………………………………………………… 37
　　3.1.2　Android 用户界面事件 ……………………………………………………………… 38
　　3.1.3　界面与布局 …………………………………………………………………………… 39
3.2　活动组件 Activity ………………………………………………………………………………… 47
　　3.2.1　AppCompatActivity、Activity 和 Context ……………………………………… 47
　　3.2.2　Activity 组件的基本方法 …………………………………………………………… 48
　　3.2.3　Activity 类具有的扩展方法 ………………………………………………………… 48
　　3.2.4　Activity 的生命周期 ………………………………………………………………… 49
　　3.2.5　手机横/竖屏自动切换问题 …………………………………………………………… 50
3.3　常用 Widget 控件的使用 ………………………………………………………………………… 51
　　3.3.1　文本控件 TextView 和 EditText …………………………………………………… 51
　　3.3.2　图像控件 ImageView ………………………………………………………………… 52
　　3.3.3　命令按钮控件 Button、ImageButton 及其单击事件监听器设计 ………………… 53
　　3.3.4　单选按钮控件 RadioButton 与复选框控件 CheckBox …………………………… 54
　　3.3.5　消息提醒对话框控件 AlertDialog 与进度控件 ProgressDialog ………………… 55
　　3.3.6　列表控件及其数据适配器和列表项选择监听器 …………………………………… 57
　　3.3.7　下拉列表控件 Spinner ……………………………………………………………… 62
3.4　高级 UI 程序设计 ………………………………………………………………………………… 63
　　3.4.1　日期和时间选择器（DatePicker 和 TimePicker）………………………………… 63
　　3.4.2　自动完成文本控件 AutoCompleteTextView ……………………………………… 64
　　3.4.3　标题栏 Toolbar 与 OptionMenu 菜单设计 ………………………………………… 64
　　3.4.4　Fragment 与 ListFragment ………………………………………………………… 66

| | 3.4.5 底部导航 BottomNavigationView | 72 |

习题 3 ··· 75

实验 3 ··· 76

第 4 章 手机基本功能程序设计 ··· 79

4.1 预备知识 ··· 79
    4.1.1 Activity 组件的调用与返回 ··· 79
    4.1.2 Android 权限、权限组与运行时权限动态检测 ·· 83
    4.1.3 SharedPreferences 存储与文件存储 ··· 87
    4.1.4 抽象类 android.net.Uri 及其静态方法 parse() ·· 92

4.2 打电话程序设计 ··· 92

4.3 短信程序设计 ·· 94
    4.3.1 SMS 简介 ··· 94
    4.3.2 短信管理器 ·· 94
    4.3.3 短信发送程序的实现 ··· 94

4.4 手机音频播放与录音程序设计 ·· 96
    4.4.1 音频播放 ·· 96
    4.4.2 手机录音 ·· 98

4.5 手机视频播放 ·· 101

4.6 手机拍照程序设计 ·· 102

习题 4 ··· 105

实验 4 ··· 106

第 5 章 服务组件及其应用 ·· 109

5.1 服务组件 Service 的基本用法 ··· 109
    5.1.1 Android 系统服务 ·· 109
    5.1.2 Service 组件及其生命周期 ··· 111
    5.1.3 自定义服务与服务注册 ·· 113
    5.1.4 服务的显式启动与隐式启动 ··· 113
    5.1.5 绑定服务方式与服务代理 ··· 115

5.2 远程服务 ··· 117
    5.2.1 远程服务概念 ··· 117
    5.2.2 Android 跨进程调用与接口定义语言 AIDL ······································· 118
    5.2.3 远程服务的建立与使用实例 ··· 118

5.3 综合应用实例——自动挂断来电后回复短信 ··· 123

习题 5 ··· 131

实验 5 ··· 132

· VII ·

第6章 广播组件与通知 ·· 135

6.1 广播与 BroadcastReceiver 组件 ·· 135
6.1.1 Android 广播机制 ··· 135
6.1.2 使用 BroadcastReceiver 组件定义广播接收者 ························· 136
6.1.3 接收系统短信广播应用实例 ·· 140

6.2 自定义广播及其使用 ··· 141
6.2.1 自定义广播 ··· 141
6.2.2 以动态注册方式使用自定义广播 ····································· 142
6.2.3 以静态注册方式使用自定义广播 ····································· 143

6.3 通知 ··· 144
6.3.1 通知与通知类 Notification ·· 144
6.3.2 通知管理器类 NotificationManager ··································· 145
6.3.3 使用 PendingIntent 查看通知内容 ···································· 145

习题 6 ··· 148
实验 6 ··· 149

第7章 SQLite 数据库编程 ··· 151

7.1 SQLite 数据库简介 ·· 151
7.1.1 SQLite 数据库软件的特点 ·· 151
7.1.2 Android 系统对 SQLite 数据库的支持 ································ 151

7.2 使用 SQLiteOpenHelper 创建、打开或更新数据库 ······················· 152
7.2.1 SQLite 数据库及表的创建与打开 ···································· 152
7.2.2 使用 SQLiteSpy 验证创建的数据库 ·································· 153
7.2.3 SQLite 数据库的更新 ·· 154

7.3 使用 SQLiteDatabase 实现数据库表的增加、删除、修改和查询 ······· 155
7.3.1 记录的增加、删除、修改和查询 ····································· 155
7.3.2 使用适配器 SimpleAdapter 显示查询结果 ··························· 157
7.3.3 以 DAO 方式访问数据库编写程序 ··································· 158
7.3.4 使用数据库事务 ··· 166

习题 7 ··· 169
实验 7 ··· 171

第8章 Android 内容提供者组件 ·· 173

8.1 ContentProvider 组件及其相关类 ·· 173
8.1.1 抽象类 ContentProvider（内容提供者） ····························· 173
8.1.2 抽象类 ContentResolver（内容解析器） ····························· 175
8.1.3 内容提供者的 Uri 定义及其相关类（UriMatcher 和 ContentUris） ·· 176

8.2 自定义 ContentProvider 及其使用 ··· 177

	8.2.1	在 Android 应用里创建并注册内容提供者	177

    8.2.2 在另一个应用程序里使用内容提供者 180

8.3 读取手机联系人信息 183

    8.3.1 手机联系人相关类 ContactsContract 183

    8.3.2 手机联系人数据库及其相关表 183

    8.3.3 读取手机联系人程序设计 184

8.4 Android 后台线程与 Android 组件的综合应用 187

    8.4.1 Android UI 主线程 187

    8.4.2 使用 Handler 向 UI 线程传递消息 187

    8.4.3 使用 AsyncTask 更新 UI 线程 189

    8.4.4 使用 ContentProvider+AsyncTask 实现群发短信 191

习题 8 199

实验 8 200

第 9 章 Android 近距离通信技术 202

9.1 WiFi 通信 202

    9.1.1 WiFi 简介 202

    9.1.2 Android 对 WiFi 的支持 202

    9.1.3 WiFi 应用实例 204

9.2 蓝牙通信 Bluetooth 208

    9.2.1 Bluetooth 简介 208

    9.2.2 Android 对 Bluetooth 的支持 208

    9.2.3 蓝牙聊天实例 211

9.3 近场通信 NFC 229

    9.3.1 NFC 简介 229

    9.3.2 Android 对 NFC 的支持 231

    9.3.3 NFC 应用实例：读/写 Tag 标签 231

习题 9 238

实验 9 239

第 10 章 位置服务与地图应用开发 241

10.1 位置服务概述 241

    10.1.1 基于位置的服务 LBS 241

    10.1.2 Android API 提供的位置包 242

    10.1.3 Google Map APIs 243

10.2 Android 定位实现 244

    10.2.1 GPS 定位实现 245

    10.2.2 网络连接及状态相关类 247

  10.2.3 WiFi 或 GPRS 定位实现 ································· 247
 10.3 百度定位及地图应用开发 ····································· 254
  10.3.1 百度定位应用开发基础 ································· 254
  10.3.2 注册百度开发者账号，申请位置应用的 Key ················· 256
  10.3.3 在清单文件中注册权限、服务及应用的 Key ················· 258
  10.3.4 百度综合定位实现 ····································· 259
  10.3.5 百度地图显示 ··· 261
 习题 10 ··· 267
 实验 10 ··· 268

第 11 章　Android 网络编程 ·· 270
 11.1 基于 HTTP 协议的 Android 网络编程 ···························· 270
  11.1.1 Android 网络编程概述 ·································· 270
  11.1.2 HTTP 请求与响应 ····································· 271
  11.1.3 使用 HttpURLConnection 访问网络资源 ····················· 271
  11.1.4 使用网络接口 HttpClient 调用 Web 服务 ···················· 274
 11.2 Android 网络图像下载与通信框架 ······························· 276
  11.2.1 网络图像下载框架 Glide ································ 276
  11.2.2 网络通信框架 Volley ··································· 277
 11.3 手机 App 与 Web 服务器通信 ·································· 283
  11.3.1 Web 服务器项目 ······································ 283
  11.3.2 App 的登录程序设计 ··································· 285
  11.3.3 App 的主界面程序设计 ································· 293
 习题 11 ··· 295
 实验 11 ··· 296
习题答案 ··· 298
参考文献 ··· 302

第 1 章
Android 应用开发概述及开发环境

随着 3G 智能手机时代的到来，人们对 Android 应用开发的需求日趋增多。Android 作为智能手机的操作系统，是新一代基于 Linux 的开源手机操作系统。手机应用软件的开发方式和环境与传统的 Windows 应用程序或者 Web 程序都有很大的不同。本章学习要点如下：

- 掌握 Android 系统的软件架构；
- 掌握搭建 Android 应用开发环境的方法；
- 掌握 Android SDK Manager 的作用；
- 掌握 Android 模拟器的使用；
- 掌握部署 Android 应用到 Android 设备中运行的方法。

1.1 移动开发与智能手机

1.1.1 移动开发概述

Android 一词最早出现于法国作家利尔亚当于 1886 年发表的科幻小说《未来夏娃》中，将外表像人的机器命名为 Android，它是一个全身绿色的机器人。

作为手机操作系统的 Android，是由安卓之父安迪·罗宾（Andy Rubin）研发完成的。Google 公司于 2005 年收购了原 Android 公司，并于 2007 年 11 月发布了基于 Linux 的开源手机平台。

最早的手机使用模拟信号，主要作用是移动电话。后来的手机使用数字信号，不仅具有移动电话功能，还可以发送短信。

1.1.2 Android 智能手机的使用特点

Android 智能手机，除了具备模拟手机打电话、发短信、蓝牙、上网等基本功能外，还具有用户定制操作系统的功能，可以像普通的计算机一样，安装或卸载应用程序。

智能手机本质上也是一台计算机，但与普通计算机有一定的差别。普通计算机的键盘、鼠标对应于较多的操作（如翻页、双击等），而手机支持各种手势对应的事件（如长

按等)。智能手机与计算机的差别如下:
- 手机只有用于返回桌面的 Home 键和退出主界面或返回到上一级界面的返回键;
- 手机的用户操作可分为按键和触屏两种。触屏事件(如滑屏、长按等)是 Android 所特有的;
- 手机进入文本编辑时,使用的是软键盘(不同于普通计算机);
- 手机系统集成了众多的硬件,如摄像头、录音机、GPS 芯片、蓝牙芯片、WiFi 网卡等。

手机的存储系统分为运行内存、手机内存和扩展存储三部分。其中,手机内存主要指系统区(包括最底层的 Linux 系统、自带的应用程序和用户应用程序)。此外,手机厂商通常会从手机内存中划分一部分存储用户数据(如照片、音乐等),即标准 SD 卡。

Android 手机的软件系统包括操作系统、中间件和一些主要应用,是基于 Java 系统,运行在 Linux 2.6 内核上的。此外,Android 手机还具有如下特点:
- Android SDK 提供多种开发所必需的工具与 API,如提供访问硬件的 API 函数,简化了摄像头、GPS 等硬件的访问过程;
- 具有自己的运行环境和虚拟机 Dalvik;
- 提供丰富的界面控件功能,加快用户界面的开发速度,保证 Android 平台上程序界面的一致性;
- 提供轻量级的进程间通信机制 Intent,使跨进程组件通信和发送系统级广播成为可能,提供了 Service 作为无用户界面、长时间后台运行的组件;
- 支持高效、快速的数据存储方式。

1.1.3 手机智能操作系统及分类

早期的手机没有操作系统 OS,内部所有的软件都是由生产商在设计时定制的,手机在设计完成后基本没有扩展功能。

为了提高手机的可扩展性,很多手机都使用了专为移动设备开发的操作系统 OS,使用者可根据需要安装不同类型的软件。

智能手机制造商所使用的半导体芯片并不都是相同的,不同的手机所采用的操作系统也可能不同。目前,主流的手机操作系统有如下几种。
- iPhone OS:由苹果公司开发的手机操作系统。
- Android:由 Google 公司发布的基于 Linux 的开源手机平台。
- Symbian:由 Symbian 公司开发和维护,后被诺基亚公司收购。该操作系统不开源。

注意:Android 手机应用开发是移动开发的一种。

1.2 Android 系统架构

Android 是基于 Linux 内核的软件平台和操作系统,采用了软件堆栈架构。该架构分为四层，即应用程序层、应用程序框架层、Android 运行时环境层和 Linux 内核层，如图 1.2.1 所示。

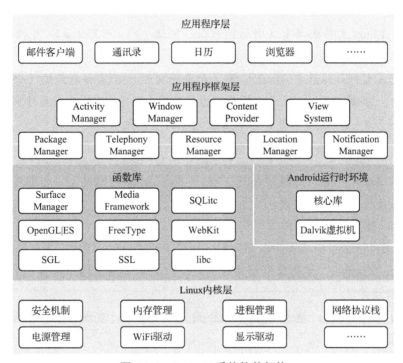

图 1.2.1　Android 系统软件架构

1.2.1　Linux 内核层

Linux 内核是硬件和其他软件堆层之间的一个抽象隔离层，提供由操作系统内核管理的底层基础功能，主要有安全机制、内存管理、进程管理、网络协议栈和驱动程序等。

Android 内核挂载/nfsroot/Androidfs 之后，根据 init.rc 和 init.goldfish.rc 进行初始化并装载系统库、程序等操作直到开机完成。init.rc 包括了文件系统初始化、装载的许多过程。init.rc 的主要工作如下：

- 设置环境变量；
- 创建 system、sdcard、data、cache 等目录；
- 把文件系统 mount 到目录，如 mount tmpfs tmpfs /sqlite_stmt_journals；
- 设置文件的用户群组、权限；
- 设置线程参数；
- 设置 TCP 缓存大小。

Android 源码编译后可得到 system.img、ramdisk.img 和 userdata.img 映像文件。其中，ramdisk.img 是 emulator 的文件系统，system.img 包括主要的包、库等文件，userdata.img 包括一些用户数据。emulator 加载这 3 个映像文件后，会把 system 和 userdata 分别加载到 ramdisk 文件系统中的 system 和 userdata 目录中。

注意：

（1）Android 手机内部存储的文件系统分区格式与 Linux 系统相同，而外部存储 SD 卡则采用 FAT。

（2）每个 Linux 文件都具有 4 种访问权限：可读（r）、可写（w）、可执行（x）和无权限（-）。

（3）目录或文件权限共有 10 位，第 1 位表示文件、目录和超链接（分别用 -、d 和小写字母 l 表示）；第 2~4 位表示文件所有者的权限；第 5~7 位表示文件所有者所属组成员的权限；第 8~10 位表示所有者所属组之外的用户权限。

1.2.2　函数库和 Android 运行时环境层

函数库和 Android 运行时环境是第二层，位于 Linux 内核之上，也称中间层，由函数库和 Android 运行时环境构成。

由于 Linux 操作系统的内核使用及其组件使用 C 语言编写（少部分使用汇编语言），因此，开发人员可以通过应用程序框架调用一组基于 C/C++ 的函数库，主要包括以下几个。

- Surface Manager：支持显示子系统的访问，为多个应用程序提供 2D、3D 图像层的平滑连接。
- Media Framework：基于 OpenCORE 的多媒体框架，实现音频、视频的播放与录制功能。
- SQLite：关系型数据库引擎。
- OpenGL | ES：基于硬件的 3D 图像加速。
- FreeType：位图与矢量字体渲染。
- WebKit：Web 浏览器引擎。
- SSL：数据加密与安全传输的函数库。
- libc：标准 C 运行库。它是 Linux 系统中底层的应用程序开发接口。

Android 运行时环境由核心库和 Dalvik 虚拟机构成。核心库为开发人员提供了 Android 系统的特有函数功能和 Java 语言的基本函数功能；Dalvik 虚拟机采用适合内存和处理器受限的专用格式。

1.2.3　应用程序框架层

应用程序框架层提供了 Android 平台的管理功能和组件重用机制，包括活动管理器（Activity Manager）、窗口管理器（Window Manager）、内容提供者（Content Provider）、视图系统（View System）、包管理器（Package Manager）、通信管理器（Telephony Manager）、资源管理器（Resource Manager）、位置管理器（Location Manager）和通知管理器（Notification Manager）。Android 的三大核心功能如下。

（1）View System：提供绘制图形，处理触摸、按键事件等功能。

（2）Activity Manager：提供管理所有应用程序的 Activity 功能。

（3）Window Manager：提供为所有应用程序分配窗口，并管理这些窗口的功能。

1.2.4 应用程序层

应用程序层提供了一系列核心应用程序，如浏览器、通讯录、相册、地图和电子市场等。

1.3 Android Studio 开发环境

1.3.1 Android Studio 概述

Android Studio 是一项全新的基于 IntelliJ IDEA 的 Android 集成开发和调试环境，与 Eclipse Android 环境相比，具有如下优点：
- Intellij IDEA / Android Studio 的智能提示很强大；
- Android Studio 内置终端，方便以命令行方式操作；
- 布局代码与效果的实时（同步）预览；
- 软件版本的联机更新；
- 项目基于 Gradle 的构建支持；
- 不仅提供了大量的组合键，还有众多快捷的设计工具、选择卡。

使用 Android Studio 开发，推荐的计算机硬件配置要求如下：
- Intel i5 以上处理器；
- 8G 及以上内存；
- 128G 及以上固态硬盘（具有较快的启动速度）。

1.3.2 下载和安装 Android Studio 3.1.2

访问 Android Studio 中文社区 http://www.android-studio.org，可以找到 Android Studio 3.1.2 的下载链接。Android Studio 3.1.2 安装分为两个阶段，首先安装 IDE（Android Studio），然后再安装 Android SDK。当 IDE 安装完成后，出现的对话框如图 1.3.1 所示。

图 1.3.1 Android Studio 首次运行

注意：

（1）Android Studio 相当于 Java 或 Java Web 开发中使用的 eclipse。
（2）Android SDK 提供了 Android 开发的软件包。
（3）Android Studio 版本在不断的更新中，而 eclipse Android 已经停止了更新。

单击 Cancel 按钮，将出现 Android SDK 的安装向导，如图 1.3.2 所示。

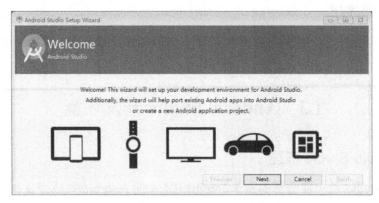

图 1.3.2　Android Studio SDK 安装向导

Android Studio 安装完成后，首次打开时呈现的主菜单，如图 1.3.3 所示。

图 1.3.3　Android Studio 主菜单

注意：打开 Android Studio 时，将会自动打开最后一次创建的项目。

1.3.3　Android Studio 相关文件夹

Android Studio 安装完成后，自动创建的几个主要文件夹如下。
- AS 安装位置 C:\Program Files\Android\Android Studio。
- AS 自带 JRE(1.8) C:\Program Files\Android\Android Studio\jre。
- Android SDK C:\Users\×××\AppData\Local\Android\sdk。
- 新建 Android 项目的保存位置 C:\Users\×××\AndroidStudioProjects，其中，×××为 Windows 安装时设定的用户名。

1.3.4　Android SDK 与 Android API

1. Android SDK

Android SDK 是 Android 软件开发包（Software Development Kit），它提供了在 Windows/

Linux/MAC 平台上开发 Android 应用的组件和各种工具集。工具集不仅包括了 Android 模拟器，还有用来调试、打包和在 Android 设备上安装应用的工具。

在 Android Studio 主菜单中，单击 Configure→Project Defaults→Project Structure，出现的对话框包含了 Android SDK 的位置和内置 JDK 的位置信息，如图 1.3.4 所示。

图 1.3.4　Android Studio 配置信息（1）

在 Android Studio 主菜单中，单击 Configure→SDK Manager，出现管理 SDK 平台及工具更新（主要是下载和卸载）的对话框，如图 1.3.5 所示。

图 1.3.5　Android Studio 配置信息（2）

注意：

（1）在 Android Studio 中打开某个项目后，其工具栏包含了 SDK Manager 工具 。

（2）在 Android Studio 中如果不关联 Android SDK，则无法开发 Android 应用程序。

在 Android SDK 文件夹里，文件夹 platforms 是主体，它是各版本开发组件的集合，包括 android.jar、字体、res 资源、模板等内容。其中，android.jar 文件提供了用于开发 Android 应用程序的编程接口（API），如图 1.3.6 所示。

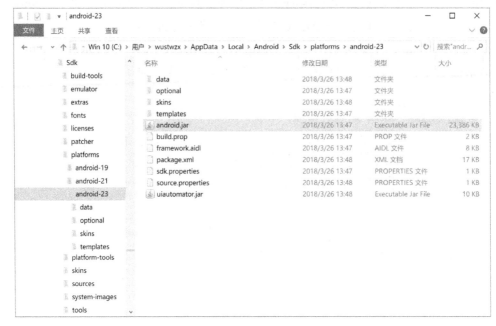

图 1.3.6　Android SDK 目录结构

除了 platforms 文件夹以外，SDK 文件夹还包含以下子文件夹。
- add-ons：存放 Android 的扩展库，如用于地图开发的 Google Maps。
- build-tools：包含各版本的 SDK 编译工具，如将.class 字节码文件转换成 Android 字节码.dex 文件的批处理程序 dx.bat、生成 Android 设备进程间通信代码的应用程序 aidl.exe 等。
- extras：扩展开发包，如 HAXM 加速。
- platform-tools：包含各版本的 SDK 通用工具，如用于将 Android 手机连接到 PC 端的 Android 调试桥（Android Debug Bridge，ADB）程序 adb.exe。又如数据库工具 SQLite。
- system-images：AVD 模拟器映像文件。
- sources：存放 Android API 的源码。
- tools：包含 avdmanager.bat、sdkmanager.bat 和 emulator.exe 等重要工具。

注意：

（1）只有下载 Android 扩展库后，才会生成文件夹 add-ons。

（2）设置 Android 应用的图标时，可以使用系统自带的图标库，这些图标文件就位于文件夹 sdk\platforms\android-19\data\res\drawable-hdpi 里。

（3）为了方便在命令行方式下使用 adb 命令，一般应将文件夹 platform-tools 的路径添加到系统环境变量 path 里。

2．Android API 核心包

标准的 Android API 包含在许多软件包里，而这些软件包又包含在文件 android.jar 里。下面介绍 Android 开发中常用的软件包。

- android.util：包含一些辅助类，如时间、日期的操作。
- android.text：包含文本处理类。
- android.text.method：提供为各种控件输入文本的类。
- android.os：提供基本的操作服务、消息传递和进程间通信，提供了 Binder、Handler、FileObserver、Looper 和 PowerManager 等类。
- android.app：实现 Android 的应用程序模型，主要包含 Activity 和 Service 组件，另外还有对话框和通知等重要类。
- android.view：提供基础的用户界面接口框架，是 Android 的核心框架，包含类 Menu、View、ViewGroup 及一系列监听器和回调函数。
- android.widget：包含在应用程序屏幕中使用的各种 UI 元素，通常派生自 View 类，包括 TextView、EditText、ImageView、ListView 和 Button 等控件。
- android.webkit：默认浏览器操作接口，包含表示 Web 浏览器的类，主要有 WebView、CacheManager 和 CookieManager。
- android.content：包含 ContentProvider 组件，还有 Context 和 Intent 等重要类。
- android.content.pm：实现与包管理器相关的类。包管理器包含各种权限、安装包、安装程序、安装服务、安装组件（如 Activity）和安装应用程序。
- android.content.res：用于访问结构化和非结构化资源文件。主要的类包括 AssetManager（用于结构化资源）和 Resources。
- android.database：实现抽象数据库的理念，提供了 Cursor 接口。
- android.database.sqlite：将 SQLite 用于物理数据库，主要包括 SQLiteOpenHelpert、SQLiteDatabase 等类。
- android.provider：提供一些类，访问 Android 的 ContentProvider，如 Contacts、MediaStore、Browser 和 Settings 等。
- android.media：提供一些类，管理多种音频、视频的媒体接口，包含 MediaPlayer、MediaRecorder、Ringtone、AudioManager 和 FaceDetector。
- android.hardware：实现与物理照相机相关的类。android.graphics.Camera 表示一种图形概念，与物理照相机完全无关。
- android.bluetooth：提供一些类来处理蓝牙功能。主要的类包括 BluetoothAdapter、BluetoothDevice、BluetoothSocket、BluetoothServerSocket 和 BluetoothClass。
- android.net：提供帮助网络访问的类，实现基本的套接字级网络 API。
- android.net.wifi：管理 WiFi 连接。
- android.telephony：提供手机设备的通话接口，包含类 CellLocation、PhoneNumberUtils

和 TelephonyManager。
- android.telephony.gsm：可用于根据基站收集手机位置，还包含负责处理 SMS 消息的类。
- android.location：定位相关类。
- com.google.android.maps：包含类 MapView 等 Google 地图所需类。
- android.gesture：包含处理用户定义的手势所需的所有类和接口。
- android.graphics：底层的图形库，包含画布、颜色过滤、点、矩形等。
- android.graphics.drawable：实现绘制协议和背景图像，支持可绘制对象动画。
- android.graphics.drawable.shapes：实现各种形状。
- android.view.animation：提供对补间动画的支持。
- android.opengl：提供 OpenGL 的工具，可 3D 加速。

1.3.5 Android Studio 常用组合键

Android Studio 开发时，为了提高编辑效率，需要掌握的快捷操作如下。

- Alt+Enter：在出现红色波浪线的地方使用，提供了许多问题的解决方案，如自动导包、自动生成接口方法和 try…catch 块等。
- Ctrl+Alt+O：优化已导入的包，清除不必要的包。
- Ctrl+D：复制光标所在行的代码至下一行。
- Ctrl+Shift+/：用于代码的注释和取消。
- Ctrl+Y：删除光标所在的一行。
- 菜单 Code→Reformat Code：代码格式化（Ctrl+Alt+L 与 QQ 快捷键冲突）。
- Ctrl+F12：显示类成员和继承的方法。
- Ctrl+H：打开类（或接口）继承关系图。
- Ctrl+O：显示所有可以重写的父类方法（含接口方法）。
- Ctrl+Alt+T：把选中的一组代码包在一块内，如 if、for 和 try…catch 等。
- Alt+Insert（MAC 无 Insert 键，可使用右键菜单→Generate）：生成代码，如 set/get 方法、构造方法等。

1.4 创建和运行 Android 应用

1.4.1 创建一个 Hello 级 Android Studio 项目

在 Android Studio 主菜单中，选择 Start a new Android Studio project 后，出现创建 Android 项目对话框。输入应用名称为"HelloAndroid"后的效果，如图 1.4.1 所示。

第 1 章　Android 应用开发概述及开发环境

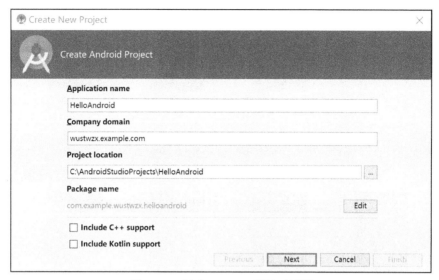

图 1.4.1　新建 Android 项目对话框

进入选择目标设备对话框，一般默认勾选 Phone and Tablet 项（表示手机和平板），根据需要指定目标设备运行的最低 Android 版本，如图 1.4.2 所示。

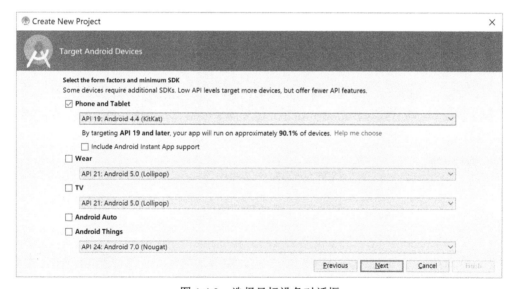

图 1.4.2　选择目标设备对话框

进入选择 Activity 模板对话框，一般默认选择 Empty Activity 项，如图 1.4.3 所示。

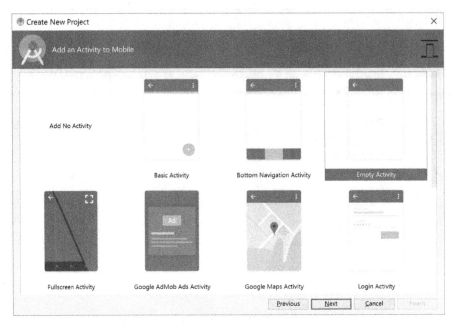

图 1.4.3 选择目标设备对话框

最后一个对话框（Login Activity）是指定应用主界面对应的 Activity 及布局名称，一般选择默认值，如图 1.4.4 所示。

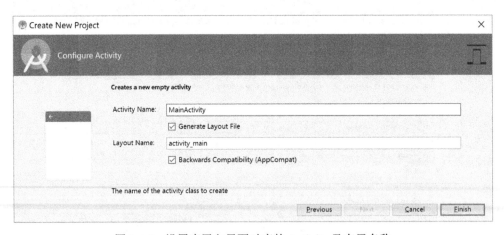

图 1.4.4 设置应用主界面对应的 Activity 及布局名称

单击 Finish 按钮，需要等待系统构建项目完成后，才能进入 Android Studio 集成开发环境，如图 1.4.5 所示。

第 1 章　Android 应用开发概述及开发环境

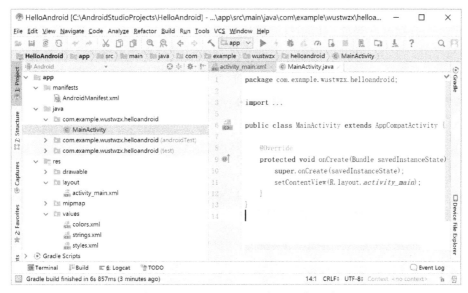

图 1.4.5　Android Studio 集成开发环境

在 Android Studio 集成开发环境里，包含菜单栏、工具栏、项目区、文档编辑区和一些选项卡。

项目结构默认使用 Android 视图，为了获取更多的信息，可以切换至 Project 视图，查看 Android API 和项目添加的依赖包。

窗口底部包含了 Terminal、Build 和 Logcat 等选项卡。其中，Terminal 用于在命令行方式下执行 Android 平台提供的一些命令；Build 用于查看项目的编译和构建信息；Logcat 用于查看 Android 设备运行的日志信息。

例如，在 Terminal 控制台，查看已连接可用的 Android 设备的命令如下：

adb devices

工具栏 用于项目 Gradle 构建，其结果在 Build 控制台里查看。如果有错误，会输出相应的信息以便修改。只有在项目构建没有错误后，方可安装和运行。

窗口左侧包含了 Project 和 Structure 两个重要的选项，默认为 Project。当编辑窗口打开某个 Java 类文件（含 Android API 的源码）时，选择 Structure 选项，就能查看该类的组成结构图。

如果已经打开某个项目窗口，再通过菜单 File→Open 打开项目，默认会出现如图 1.4.6 所示的对话框。

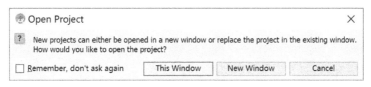

图 1.4.6　Android Studio 默认打开项目方式

一旦选择 This Window 并勾选了左边的任选框，以后再打开项目时就不会出现上述对话框了，而是默认在当前窗口中打开其他项目。

在 Android Studio 开发过程中，有时需要同时打开多个项目窗口。解决办法是使用菜单 File→Settings→Appearance & Behavior→System Settings→Project Opening，选中 Confirm window to open project in 项，如图 1.4.7 所示。

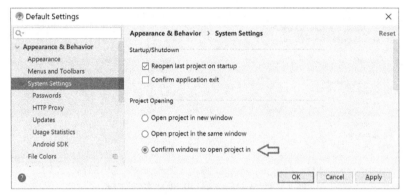

图 1.4.7　Android Studio 设置允许打开多个项目窗口

注意：

（1）不同 Android Studio 版本的界面布局有一定的差异，本书使用的是 3.1.2 版本。

（2）项目创建时指定的目标设备最低 Android 版本信息并未保存在项目清单文件 AndroidManifest.xml 里。

（3）打开 Java 类文件时，通过【Ctrl+Click】组合键方式可查看 Android API 及其源码。

（4）在编辑某个项目的同时，使用系统菜单 File 也可以新建 Android 项目。

1.4.2　手机模拟器与 AVD Manager

1．使用 Android Studio 内置模拟器

AVD（Android Virtual Device）是 Android SDK 提供的最重要工具之一，它使开发人员在没有物理设备的情况下，可以在计算机上对 Android 程序进行开发、调试和仿真。

Android Studio 工具栏上的 AVD Manager 工具，用于创建、编辑和运行模拟器，如图 1.4.8 所示。

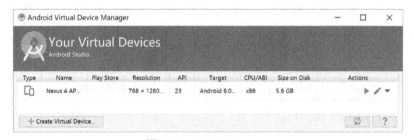

图 1.4.8　AVD Manager

第1章　Android应用开发概述及开发环境

在 Android Studio 中，可以创建和启动多个模拟器，每个模拟器对应一个 ID，它就是模拟器的电话号码。其中，第一个启动的模拟器 ID 为 5554，第二个启动的模拟器 ID 为 5556，Android 系统允许手机（或模拟器）向自己发送短信息。

注意：Android Studio 已经不再支持在模拟器控制(Emulator Control)中操作模拟器，改为在 Android 模拟器上进行操作。单击模拟器右侧下方的三个小点，在弹出的 Extended controls 对话框中进行操作，如在 Phone 选项卡里，可模拟打电话和发短信操作。

2．使用第三方模拟器

Android 内置模拟器启动速度较慢，操作也不够流畅。使用 Android 手机进行测试，可以弥补上述不足，但却连接不便。因此，业界也开始广泛使用第三方模拟器。下面介绍一款对硬件要求较低、基于 Android 4.4 的夜神安卓模拟器。

访问课程网站 http://www.wustwzx.com/android/index.html，可下载夜神安卓模拟器安装包。为了部署 Android 应用到该模拟器，在启动该模拟器后，进入 Android Studio 的 Terminal 终端，执行桥接命令 adb，如图 1.4.9 所示。

图 1.4.9　创建 Android Studio 与夜神 AVD 的连接

3．设备文件浏览器

在 Android Studio 3.X 里，使用菜单 View→Tool Windows→Device File Explorer，可浏览 Android 设备里的文件，实现文件管理工作。既可以查看文件夹 data 里用户安装的应用；也可以查看文件夹 sdcard，包括文件的打开、删除、导入与导出等操作，如图 1.4.10 所示。

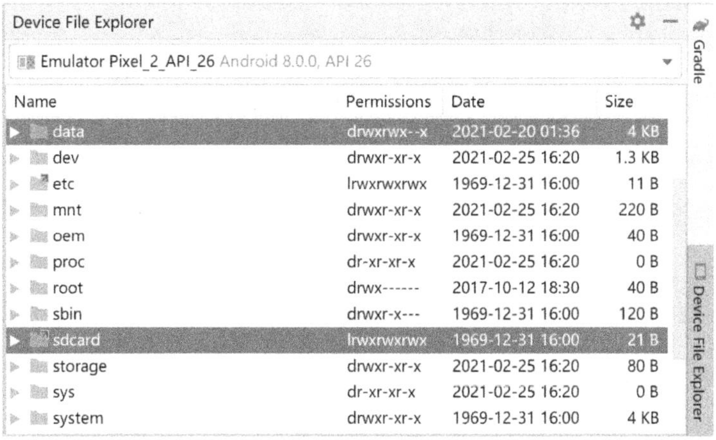

图 1.4.10　Android Studio 的设备文件浏览窗口

15

注意：

（1）Android Studio 界面的右下方，提供了快速进入设备文件浏览器的工具 Device File Explorer。

（2）文件夹 data 属于内部存储，而 sdcard 属于外部存储。在文件夹 sdcard 里，分别存放了照片、音乐和视频等子文件夹。

（3）对于没有 root 权限的 Android 手机，无法访问某些位于手机内部存储的系统文件夹（如位于/data/app/packname 里的.apk 文件）；而模拟器里的所有文件均可被访问。

1.4.3 安装和运行 Android 应用

安装 Android 应用到 Android 手机上运行之前，通常需要将手机与计算机相连接。在物理上将手机通过手机数据线与计算机连接前，应打开手机的 USB 调试开关，其方法是运行手机的设置→开发人员选项→USB 调试。

单击 Android Studio 工具栏上的 ▶ 图标，弹出 Select Deployment Target 对话框，选择对应的手机或虚拟机设备，即可安装和运行。

注意： 手机较 AVD 而言，响应速度更快。

如果 Android 设备的版本低于建立 Android 应用项目时指定的最低版本，则无法安装应用。解决办法：单击 Android Studio 工具栏上的 工具，依次单击 App 项和 Flavors 项，更改 Min Sdk Version 选项值，如图 1.4.11 所示。

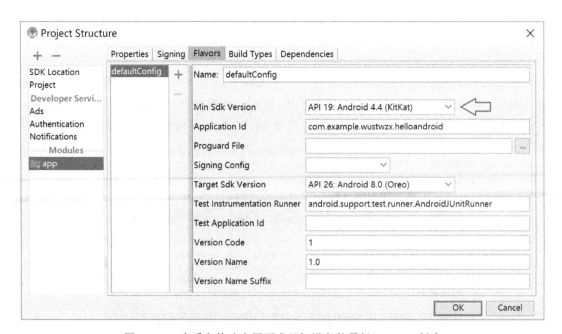

图 1.4.11　查看和修改应用要求目标设备的最低 Android 版本

习 题 1

一、判断题

1．使用手机自带的文件管理程序，可以浏览 Android 应用的安装目录。
2．在计算机上安装手机驱动程序之前，必须打开手机的 USB 调试开关。
3．在安装 Android Studio 之前，必须安装 JDK。
4．在 Android Studio 中，SDK Manager 的主要作用是 Android 平台的下载管理。
5．在 Android Studio 创建项目的对话框中，包含了对运行设备最低 Android 版本的指定。
6．AVD 是 Android 虚拟设备的英文缩写。

二、选择题

1．下列不属于 Android 应用程序框架层的是_____。
   A．Activity Manager　　　　　　B．SQLite
   C．Location Manager　　　　　　D．Notification Manager
2．Android 移动设备底层（内核）使用_____系统。
   A．DOS　　　　B．Linux　　　　C．Windows　　　D．UNIX
3．用于 Android 应用开发的 android.jar 文件，位于 sdk 文件夹的_____子文件夹里。
   A．add-ons　　　　　　　　　　B．platforms
   C．system-images　　　　　　　D．tools
4．在 Android Studio 创建 AVD 时，默认选择的目标设备是_____。
   A．TV　　　　　　　　　　　　B．Android Auto
   C．Phone and Tablet　　　　　　D．Wear
5．按照缩进风格格式化程序代码，应使用 Android Studio 的_____菜单。
   A．Build　　　B．Tools　　　C．View　　　D．Code

三、填空题

1．通常将 Android 软件系统划分为_____层。
2．在 Android 系统中，应用程序设计使用_____语言。
3．在 Android Studio 中，使用组合键_____优化程序所需要的软件包。
4．在 Android Studio 中，先启动的内置模拟器的电话号码是_____。
5．使用 Android Studio 3.1.2 提供的设备文件浏览器功能,应选择的菜单是_____。

实 验 1

一、实验目的

（1）掌握 Android 集成开发环境的搭建方法。
（2）掌握 Android SDK 的作用及 Android SDK Manager 的使用。
（3）掌握使用向导创建 Android Hello 项目的方法。
（4）掌握 Android 模拟器的创建与使用。
（5）掌握 Android 项目的部署及运行方法。

二、实验内容及步骤

1. 快速搭建 Android 集成开发环境

（1）访问 Android 中文社区 http://www.android-studio.org，依次选择下载→历史版本下载，下载 Android Studio 3.1.2 版本后进行安装。
（2）Android Studio 安装完成后，再安装 Android SDK。
（3）在 Android Studio 中，使用 Android SDK Manager 工具 下载所需要的 Android API 版本（如 API 25，对应于 Android 7.1.1）。

2. 使用向导，创建 Android Studio 项目

（1）在 Android Studio 中，使用菜单 File→New→New Project。
（2）在出现的对话框中，全部选择默认值。
（3）使用 Android Studio 工具栏上的 工具，查看项目的配置信息。
（4）使用【Ctrl+Click】组合键方式，查看 Android API 及其源码。
（5）逐步掌握 Android Studio 快捷键的使用。

3. Android 模拟器的使用

（1）单击 Android Studio 工具栏上的模拟器工具 ，进入模拟器管理界面。
（2）创建一个与已有 API 版本相对应的模拟器。
（3）启动模拟器，模拟手机的实际操作。
（4）单击 Android Studio 界面右侧下方的设备文件管理器工具按钮，对模拟器文件进行导入/导出操作。

4. 部署 Android 项目并做运行测试、应用程序卸载

（1）确保手机开发者选项可用，打开 USB 调试开关。
（2）单击 Android Studio 工具栏上的工具 ，分别部署项目到手机和模拟器中并运行。
（3）练习卸载 Android 应用的操作。
（4）查看 Android Studio 的各控制台（如 Build、Run 和 Logcat 等）的输出信息。

三、实验小结及思考

（由学生填写，重点写上机中遇到的问题。）

第 2 章 Android 项目结构分析及调试

Android 系统提供了 Activity、Service、BroadcastReceiver 和 ContentProvider 四大组件和用于组件通信的相关类，同时，还提供了用于通知、音频、视频、WiFi、Bluetooth、位置服务和通信等应用的管理器类。因此，Android 应用是基于组件的编程。

在通常情况下，一个 Android 应用有一个主界面，并使用 Android 的 Activity 组件来实现。本章将具体阐述 Android 应用项目的结构、组成部分和 Android 虚拟机的工作原理，学习要点如下：

- 掌握 Android 项目的文件系统结构；
- 初步了解 Android 四大组件，特别是 Activity 组件的作用；
- 初步了解 Activity 组件的相关类（如 View、Intent 等）的作用；
- 掌握 Android 项目清单文件 AndroidManifest.xml 的作用；
- 了解 Java 虚拟机的工作原理；
- 掌握 Android 项目的动态调试和单元测试方法。

2.1 Android 项目的文件系统结构

由于 Android 将用户界面和资源从业务逻辑中分出来，并使用 XML 文件进行描述形成独立的资源文件。因此，Android 应用项目的文件系统结构比 Java/Java Web 项目更为复杂，特别是资源文件的调用关系。

在 Android Studio 中，项目有多种视图查看方式。默认使用 Android 视图，创建项目时自动生成的模块 App 包含 manifests、java 和 res 三个文件夹。Android Studio 还包括 Project 和 Packages 等视图。

2.1.1 项目配置清单文件夹 manifests

每个 Android 项目都有一个名为 AndroidManifest.xml 的文件，它是 XML 格式的文件，包含了 Android 系统运行前必须掌握的相关信息，如应用程序名称、图标、应用程序的包名、组件注册信息和权限配置等，如图 2.1.1 所示。

```xml
<?xml version="1.0" encoding="utf-8"?>
<manifest xmlns:android="http://schemas.android.com/apk/res/android"
    package="com.example.example4_6">
    <uses-permission android:name="android.permission.WRITE_EXTERNAL_STORAGE"/>
    <application
        android:allowBackup="true"
        android:icon="@mipmap/ic_launcher"
        android:label="@string/app_name"
        android:roundIcon="@mipmap/ic_launcher_round"
        android:supportsRtl="true"
        android:theme="@style/AppTheme">
        <activity android:name=".MainActivity">
            <intent-filter>
                <action android:name="android.intent.action.MAIN" />
                <category android:name="android.intent.category.LAUNCHER"/>
            </intent-filter>
        </activity>
    </application>
</manifest>
```

图 2.1.1　Android Studio 项目的清单文件结构

在 Android Studio 中创建一个新项目时，Application ID 默认和项目的包名一致，即 Android 的应用 ID 与包名是绑定的。在 Android API 中，一些方法和参数从名称上看似乎返回的是包名。事实上，它们返回的是应用 ID 值。例如，Context.getPackageName() 方法返回的是应用 ID，而不是包名。

设置不同的应用 ID，就相当于划分了同一应用的不同版本，这比修改应用的包名更加方便。因为，应用的包名修改后，可能会导致程序里的其他相关修改。

2.1.2　源程序文件夹 Java

Android 以 Java 作为编程语言，因此其程序文件以.java 作为扩展名。Java 程序文件位于 src 文件夹的某个包内。

注意：

（1）包名习惯用小写，而类名首字母习惯用大写。

（2）src 文件夹里可以建立若干个包，用以分类存放 Java 源程序文件。

2.1.3　资源程序文件夹 res

在 Android 项目中，有字符串、位图、布局等资源，可以将其划分为三种类型：XML 文件、位图（图像）文件和 raw（声音）文件。

在 Android 项目中，有两个用于存放资源文件的文件夹，分别为 res 和 assets。其中，res 文件夹内的资源文件最终被打包到编译后的.java 文件中，res 文件夹内不支持深度的子目录；assets 文件夹中的资源文件不会被编译，而是直接打包到应用中，assets 文件夹支持任意深度的子目录。

Android 系统自动为每个资源分配一个十六进制的整型数，用以标明每个资源，保存在名为 R.java 的文件里（在 Packages 视图方式下可以查看），示例代码如下：

```java
package com.example.hello;
public final class R {
    public static final class attr {
    }
    public static final class dimen {
        public static final int activity_horizontal_margin=0x7f040000;
        public static final int activity_vertical_margin=0x7f040001;
    }
    public static final class drawable {
        public static final int ic_launcher=0x7f020000;
        public static final int wzx=0x7f020001;
        public static final int wzx2=0x7f020002;      }
    public static final class id {
        public static final int action_settings=0x7f080002;
        public static final int button1=0x7f080001;
        public static final int textView1=0x7f080000;      }
    public static final class layout {
        public static final int activity_main=0x7f030000;      }
    public static final class menu {
        public static final int main=0x7f070000;      }
    public static final class string {
        public static final int action_settings=0x7f050001;
        public static final int app_name=0x7f050000;
        public static final int hello_world=0x7f050002;      }
    public static final class style {
        public static final int AppBaseTheme=0x7f060000;
        public static final int AppTheme=0x7f060001;      }
}
```

注意：res 文件夹内的资源文件可以通过 R 资源类访问，而 assets 文件夹内的资源文件则不能。

1．布局文件夹 res/layout

布局文件夹 res/layout 用来存放扩展名为.xml 的布局文件，由某种布局管理器管理的若干控件对象组成，供 Activity 组件使用。

2．值文件夹 res/values

值文件夹 res/values 里的 strings.xml 是非常重要的文件，通常存放着布局文件中控件对象的属性值。

3．软件设计的国际化

国际化是指在软件设计过程中将特定语言及区域脱钩的过程。当软件移植到不同的语言及区域时，软件本身不需要做任何的修改。Android SDK 并没有提供专门的 API 来实现国际化，而是通过对不同的资源文件进行不同的命名来达到国际化的目的。

例如，在布局文件中，定义文本框控件的代码：

```
<TextView …android:text="@string/hello " />
```

就是符合国际化的做法,控件值来源于文件 res/values/strings.xml,通过键名 hello 来引用。

如果不通过引用字符串变量的方式,而是直接把字符串常量写在 TextView 控件的 android:text 属性后,其代码如下:

```
<TextView …android:text="字符串常量" />
```

这种写法,对程序运行没有任何影响,只是不符合国际化的做法。

注意:使用软件设计的国际化,能有效实现程序员与 UI 设计人员工作的分工协作。

4. 图像文件夹 res/drawable

与 Windows 应用程序一样,每个 Android 应用项目都有一个图标。Android 应用默认使用的图标文件 ic_launcher 是一个绿色的机器人,其文件格式是.xml,存放在文件夹 res/mipmap 里。更改 Android 应用默认图标的一种方法:右击 mipmap→New→Vector Asset →Clip Art。另一种方法:先将.png 格式的文件,复制到文件夹 res/drawable 中,然后通过 R 文件引用该图像文件。

5. 声音文件夹 res/raw

项目使用的音频文件,通常存放在 res/raw 文件夹里。

2.1.4 项目多模块及构建 Gradle Scripts

1. 创建多模块

在 Android Studio 开发中,有时希望把多个相关联的应用集合在一个项目里。使用 Android Studio 提供的多模块功能可以轻松地做到这一点。在一个已经创建的项目 Example5 里,创建一个新模块的操作方法:File→New→New Module,如图 2.1.2 所示。

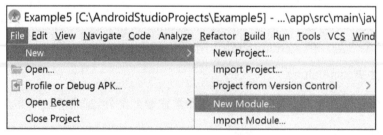

图 2.1.2 创建 New Module

创建 New Module 的方法与创建项目的步骤基本相同,只是 New Module 的名称取代了项目名称。在项目 Example5 里创建名为 example5-1 的 Module 后,项目结构如图 2.1.3 所示。

第 2 章　Android 项目结构分析及调试

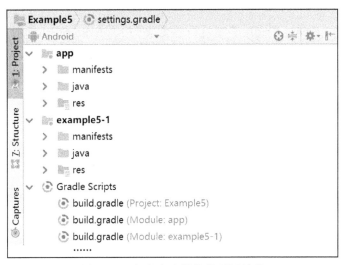

图 2.1.3　包含多个模块的项目结构

注意：

（1）创建项目时，自动生成的 app 实质上就是一个 Module，它会调用后来创建的 Module。

（2）在工具栏里，可以选择不同的 Module 单独进行调试和安装。

2．Gradle 构建 Scripts 与依赖管理

Gradle 是一种依赖管理工具，基于 Groovy 语言主要面向 Java 应用，它抛弃了基于 XML 的各种烦琐配置，取而代之的是一种基于 Groovy 的内部领域特定（DSL）语言。

一个 Android 应用中的每个 Module 都对应一个名为 build.gradle 的脚本文件，它包含了该 Module 使用的 API 版本、仓库（Repositories）和依赖（Dependencies）等。其中，依赖管理以可视的方式进行，如图 2.1.4 所示。

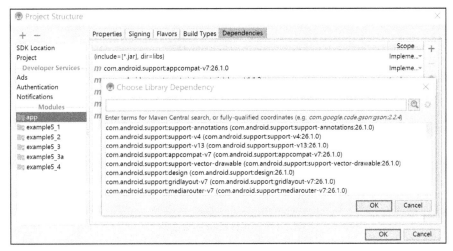

图 2.1.4　项目（模块）的依赖管理

注意：

（1）脚本文件 settings.gradle 声明有一些需要加入 gradle 的 Module。

（2）脚本文件 gradle-wrapper.properties 用来配置 Gradle 及其版本。

（3）在 Android Studio 里打开项目时，会加载其包含的 Module，这比打开多个项目窗口进行参照编辑更加方便。

（4）项目构建脚本是面向所有 Module 的，文件 settings 包含了项目的所有 Module 名称。

（5）模块调试通过后，为了添加新模块有更快的速度，可以使用模块快捷菜单里的选项 Load/Unload Modules 实现不加载。必要时，再次使用项目的快捷菜单里的该选项加载被 Unload 的模块。

2.1.5 使用 Project 或 Packages 视图

在 Android Studio 中创建的项目，默认使用 Android 视图。有时，也需要进行一些非常规的操作，如获取应用的安装包文件（*.apk）、添加第三方的依赖 jar 包等，这就需要切换至 Project 视图。快速浏览 Android API 时，可选择 Packages 视图。

1. 项目编译与生成

工具栏中的工具 用于按照 Gradle 文件同步引用库。跨语言、跨 Module 的项目一般在修改后都需要同步一下。当项目增加了音乐文件等资源时，也需要使用该工具来更新项目自动生成的资源文件 R.java（在 Packages 视图下可查看到）。

在部署项目至 Android 前，会先进行项目构建。工具栏里的 工具用于项目生成（Make Project），其结果信息显示在 Build 控制台。如果有错误会输出相应的信息，以便修改；如果没有错误会安装至 Android 设备和运行。项目成功生成后，将项目视图切换至 Project 视图，在项目文件夹 build/outputs/apk 里，可以查看到生成的安装包文件（app-debug.apk），如图 2.1.5 所示。

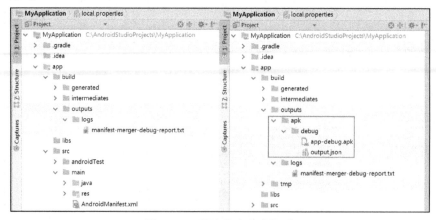

图 2.1.5　项目成功生成前后

使用菜单操作项目的要点如下。

- Make Project：编译 Project 下所有 Module，一般是自上次编译后 Project 有更新的文件。首次编译无误时会生成.apk 文件，以后编译时将不会生成.apk 文件。
- Clean Project：删除先前的编译文件，包括 apk 文件夹。
- Rebuild Project：先执行 Clean 操作，然后重新编译。

2．将项目依赖的.jar 包复制到文件夹 libs

当使用第三方库文件时，需要将其.jar 包复制到项目的 libs 文件夹里。例如，做百度位置应用开发时，就使用了百度提供的.jar 包。

2.2 Android 应用程序的基本组成

Android 应用程序是由组件组成的，组件可以调用相互独立的功能模块。根据完成的功能，组件可划分为四类核心组件，即 Activity、Service、BroadcastReceiver 和 ContentProvider。

注意：

（1）在结构上，Android 应用程序与传统的 C 语言程序不同，它是基于组件编程的。

（2）Android 四大组件中除 ContentProvider 组件外，都是通过 Intent 对象激活的。

（3）Android 四大组件，均需要在项目清单文件中使用相关标签进行注册。

2.2.1 Activity 组件与视图 View

Activity 是 Android 最重要的组件，负责用户界面的设计。Activity 用户界面框架采用 MVC 模式（Model View Controller）。控制器负责接受并响应程序的外部动作；通过视图反馈应用程序给用户的信息（UI 界面反馈）；模型是应用程序的核心，用于保存数据和代码。

注意：Activity 组件在清单文件中使用标签<activity>注册。

2.2.2 Service 组件

Service 是 Android 提供的无用户界面、长时间在后台运行的组件。Android 提供了许多系统服务程序。

注意：Service 组件在清单文件中使用标签<service>注册。

2.2.3 BroadcastReceiver 组件

在 Android 系统中，当有特定事件发生时就会产生相应的广播。例如，开机启动完成、收到短信、电池电量改变、网络状态改变等。

为了通知手机用户有事件发生，在通常情况下，通知管理器（NotificationManager）会在手机的状态栏里产生一个具有提示音的通知,用户通过下滑手势可以查看相关信息。BroadcastReceiver（广播接收者）接收来自系统或其他应用程序的广播，并做出回应。

注意：

（1）BroadcastReceiver 组件在清单文件中使用标签<receiver>注册；

（2）BroadcastReceiver 组件与 Service 组件一样，也没有 UI 界面。

2.2.4　ContentProvider 组件

为了跨进程共享数据，Android 提供了 ContentProvider 组件，可以在无须了解数据源、路径的情况下，对共享数据进行查询、添加、删除和更新等操作。

注意：

（1）ContentProvider 组件在清单文件中使用标签<provider>注册。

（2）ContentProvider 组件的使用详见第 8 章。

2.2.5　Application、Context 和 Intent

1．应用对象 Application

当 Android 程序启动时系统会创建一个 Application 类型的对象，用来存储系统的一些信息完成数据传递、共享和缓存等操作。

Application 对象的生命周期是整个程序中最长的，它的生命周期就等于这个程序的生命周期，且是全局、单例的，即在不同的 Activity 和 Service 中获得的对象都是同一个对象。

2．上下文对象 Context

Activity 和 Service 都是 Context 的子类，通过 Context 提供的方法 getApplicationContext() 就能获得 Context 对象。通过 Intent 对象，Activity 组件之意可以相互调用，实现有参数传递或返回值的调用。

BroadcastReceiver、ContentProvider 并不是 Context 的子类，其所持有的 Context 都是由其他组件传递过来的。在 Activity 和 Service 组件里发送广播，均需要使用 Intent 对象。

Android 组件及通信机制，如图 2.2.1 所示。

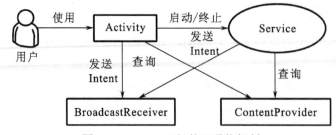

图 2.2.1　Android 组件及通信机制

3．意图对象 Intent

Android 提供轻量级的进程间通信机制 Intent，使跨进程组件通信和发送系统广播成为可能，组件 Activity、Service 和 BroadcastReceiver 都是通过消息机制被启动（激活）的，其使用的消息就封装在对象 Intent 里。

注意：Context 是一个抽象类且为 Activity 的超类，提供了 startActivity()方法，并以 Intent 对象作为参数，用于实现窗体的跳转。

在 Android 中，调用其他应用的动作名称由众多的类提供，也可以自定义。Android 系统提供的常用意图动作名称，如表 2.2.1 所示。

表 2.2.1 Android 中常用的意图动作名称

方 法 名	功 能 描 述
Intent.ACTION_MAIN	标识 Activity 为一个程序的开始,参见清单文件中对主 Activity 的定义
Intent.ACTION_DIAL	调用系统提供的拨号程序
Intent.ACTION_CALL	呼叫指定的电话
Intent.ACTION_SENDTO	发短信、E-Mail 等
Intent.ACTION_VIEW	浏览网页、地图、播放多媒体等
Intent.ACTION_WEB_SEARCH	网络搜索
Intent.ACTION_BATTERY_CHANGED	检测手机电量情况
BluetoothAdapter.ACTION_REQUEST_ENABLE	蓝牙当前是否可用

2.2.6 Android 应用程序的运行入口

在一个 Android 应用程序开始运行的时候，会单独启动一个进程（Process）。默认的情况下，这个应用程序中的所有组件（Activity、Service、BroadcastReceiver 和 ContentProvider）都会运行在这个进程里。

一个 Android 应用程序通常由多个 Activity 组成，但只有一个主 Activity。在项目清单文件中使用<activity>标签注册主 Activity 时，还需要内嵌<intent-filter>、<action>和<category>标签，以此说明该 Activity 为 Android 应用程序的入口。定义一个 MainActivity.java 为应用的主 Activity 代码如下：

```
<activity
    android:name="prg_packname.MainActivity"
    android:label="@string/app_name">
    <intent-filter>
        <action android:name="android.intent.action.MAIN"/>
        <category android:name="android.intent.category.LAUNCHER" />
    </intent-filter>
</activity>
```

其中，prg_packname 为程序 MainActivity.java 的包名。

【例 2.2.1】 调用系统提供的拨号程序。

可以将系统拨号程序理解为系统预先定义的一个 Activity，通过 Intent 对象去激活。在此之前，需要设置其动作和数据两个属性。程序运行效果如图 2.2.2 所示。

图 2.2.2　程序运行效果

设计步骤如下：
（1）在项目的布局文件里，添加一个名为 call_btn 的 Button 控件。
（2）在 MainActivity 程序的 onCreate()方法里，添加如下代码：

```
//先找控件，后设置监听器（使用匿名内部类创建监听器对象）
findViewById(R.id.call_btn).setOnClickListener(new View.OnClickListener() {
    @Override
    public void onClick(View v) {
        //创建意图对象
        Intent intent=new Intent();
        //调用系统的界面程序
        intent.setAction(Intent.ACTION_VIEW);
        //intent.setAction(Intent.ACTION_DIAL);   //调用系统的拨号程序
        //根据数据类型打开相应的 Activity（拨号程序）
        intent.setData(Uri.parse("tel:10086"));
        startActivity(intent);
    }
});
```

（3）部署应用并做运行测试。

2.3　Android 虚拟机 Dalvik

虚拟机（Virtual Machine）是通过软件模拟的具有完整硬件系统功能、运行在一个隔离环境中的计算机系统。

尽管 Android 的编程语言是 Java，但 Android 使用的虚拟机 Dalvik 与 Java 使用的虚拟机 JVM 并不兼容。因为 Dalvik 是基于寄存器的架构，而 JVM 是基于栈的架构。此外，Dalvik 能根据硬件实现更大的优化，更适合于移动设备。

Android 应用程序在编译成.class 文件后，还会通过一个批处理程序（dx.bat）将应用所有的.class 文件转换成一个名为 classes.dex 的文件。

Android 安装包文件的扩展名为.apk，一个项目只能放进一个.apk 文件内。Android 安装包文件包含了与某个 Android 项目相关的所有文件，包括项目的清单文件 AndroidManifest.xml、应用程序代码 classes.dex、资源文件和其他文件打成的一个压缩包。解压后的文件系统如图 2.3.1 所示。

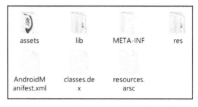

图 2.3.1 被解压后的文件系统

当用户部署一个自己开发的 Android 应用至手机内存后，在系统目录/data/app 里可以找到该应用的安装包文件（*.apk）。在系统目录/data/dalvik-cache 里可以找到对应的.odex 文件，它是从 classes.dex 文件优化而来的，或者说是从.apk 文件中提取出来的可运行文件。

运行用户开发 Android 应用程序时，如果在系统目录\data\dalvik 里存放相应的.odex 文件，Android 的 Dalvik 虚拟机会直接从.odex 文件中加载指令和数据后执行；若没有存放，则需要先从.apk 包中提取 classes.dex 并生成.odex 文件，然后再加载并执行。因为真正在 Android 虚拟机上运行的是.odex 文件，如果系统发现已经有了.odex 文件，那么就不会再从.apk 包里面去解压、提取。显然，这种预先提取方式可以加快软件的启动速度，减少对 RAM 的占用。

注意：

（1）Android 系统文件夹/system/app 存放的是系统的默认组件、相应的.apk 安装包文件和直接加载执行的.odex 文件，如图 2.3.2 所示。

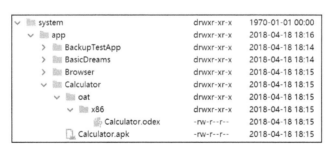

图 2.3.2 Android 系统自带的应用程序

（2）Android 系统文件夹/system/bin 用于存放 Linux 系统自带的组件，如包管理器/system/bin/pm。

2.4 Android Studio 项目调试

2.4.1 主动调试（Toast 与 Logcat）

android.widget.Toast 类用于实现消息提醒，其信息在显示几秒后自动消失。Toast 可以看作是一个会自动消失的信息框，它只能是以程序代码的方式设计。Toast 类的定义及主要方法如图 2.4.1 所示。

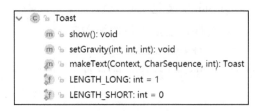

图 2.4.1　Toast 类的定义

Toast 的静态方法 makeText(Context,CharSequence,int)用于创建一个 Toast 对象。其中，第 1 个参数是上下文对象，表示在哪个 Activity 里显示；第 2 个参数是要显示的字符串信息；第 3 个参数是显示的时长，可使用类常量 Toast.LENGTH_LONG 或 Toast.LENGTH_SHORT，还可以使用以毫秒为单位的正整数。

通常，先使用 makeText()方法创建一个 Toast 对象，再使用 show()方法显示某个 Activity。在使用 show()方法前，还可使用 setGravity()方法设置其在屏幕显示的位置。

Toast 类的示例用法如下：

Toast. makeText(this, "显示信息",Toast.LENGTH_LONG).show();

注意：
（1）不使用 setGravity(int,int,int)时，默认出现在手机屏幕偏下的位置。
（2）makeText()方法里参数 this 表示当前 Activity 的上下文。
（3）输入 Toast 并按回车键，能快速产生其余的代码。

Android 系统运行应用程序时会产生一些日志（Log）信息，按照级别从低到高的顺序，它们依次划分为 Verbose（详细）、Debug（调试）、Information（信息）、Warning（警告）和 Error（错误）等不同级别。

如果 Android 程序运行时自动闪退，则表明程序遇到了致命的错误。打开 Logcat 控制台，选择日志级别为 Error，可获取错误信息，如空指针异常等。

注意：当程序自动闪退时，也可以通过查看 Run 控制台，找到出错的原因。

如果 Android 程序运行时出现逻辑错误，应该得到的结果没有得到，可在程序里安排一些日志输出，以便定位错误的位置。

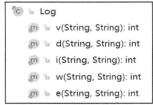

图 2.4.2　Log 类的定义

Log 是一个 Android 类，提供了产生不同级别日志信息的静态方法，如图 2.4.2 所示。

其中，第 1 个参数为用户自定义方便筛选的标签，第 2 个参数为需要输出的信息。

Logcat 是 Android 提供的一个日志浏览器，将程序运行时产生的日志信息输出到 Logcat 控制台，以便定位应用程序产生错误的位置或观察特定的运行信息，如图 2.4.3 所示。

第 2 章 Android 项目结构分析及调试

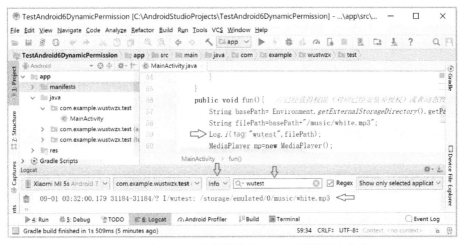

图 2.4.3 在 Logcat 控制台筛选 Info 级别的日志信息

注意：

（1）编程时，输入 logi 并按回车键能快速产生在 Logcat 控制台输出 Info 信息的语句。

（2）Logcat 控制台输出不同级别的信息，使用不同的颜色。例如，e 级的错误信息为红色。

（3）使用 Toast 和 Logcat 均属于程序员自己进行的静态调试，与下面介绍的 Debug 或测试框架完成的调试不同。

2.4.2 动态调试

如同 VC++和 Visual Studio 等开发环境，在 Android Studio 环境中，也可以通过设置断点、检查变量值的方式来检查错误，适用于程序错误的快速精准定位。

单击某行行号阴影区域即可设置断点，将产生一个断点标记，如图 2.4.4 所示。

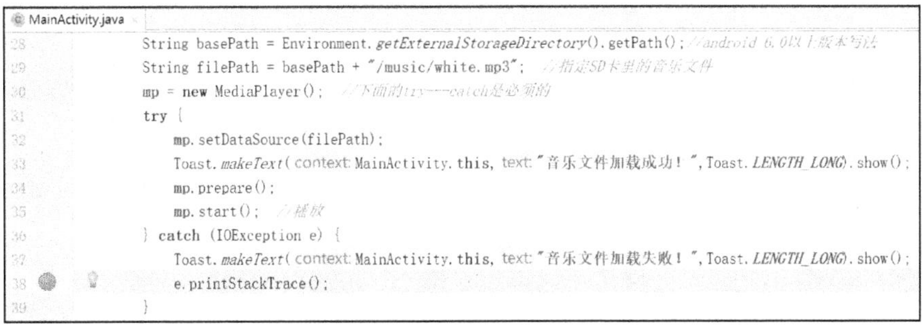

图 2.4.4 设置断点的方法

注意： 再次单击断点标记，将取消断点设置。

单击工具栏上的 图标，将以 Debug 方式运行程序。在各个断点处暂停程序运行并显示内存变量值。此时，单击 图标，可以以单步方式运行程序。通过这种调试方法，可以迅速找到问题的原因，如图 2.4.5 所示。

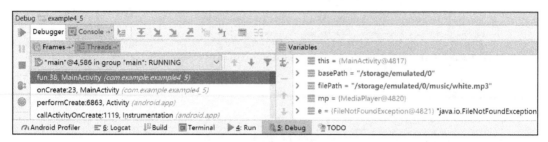

图 2.4.5 Debug 运行方式

图 2.4.5 表明，程序暂停在 38 行，变量区的 e 变量值表明前面出现了文件未找到的异常（FileNotFoundException），其原因是没有扩展存储文件的读取权限（Permission Denied）。

2.4.3 单元测试

开发一个 Android 项目时，可能需要编写很多的业务逻辑类，其正确性可以单独测试而不必在整个应用中调试。测试也是开发过程中的一个重要组成部分，Android 开发环境默认集成了 JUnit 测试框架，可以对 Android 项目进行单元测试。

【例 2.4.1】 使用 JUnit 测试一个 Android 项目中某个辅助类方法的正确性。

【设计步骤】

（1）新建名为 JUnitTest 的 Android 项目，在其对话框中均采用默认设置。

（2）在 junittest 包里新建名为 Calculator.java 的类文件。

此时，Android 项目文件结构如图 2.4.6 所示。

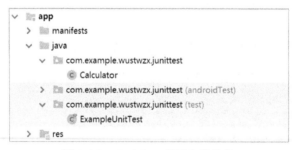

图 2.4.6 JUnitTest 项目文件结构

（3）编写被测试类 Calculator.java，其代码如下：

```
package com.example.wustwzx.junittest;
//定义被测试类，它作为 Android 项目的一个辅助类
//测试本类的方法 add()和 subtract()的正确性
public class Calculator {
    public int add(int num1, int num2){
        return num1+num2;
    }
    public int subtract(int num1, int num2){
```

```
        return num1-num2;
    }
}
```

（4）在被测试类 Calculator.java 代码区域内右击，依次选择 Go to → Test，进入 Create Test 界面。

（5）在 Member 区域选择需要自动生成的测试方法，在 add()与 subtract()方法前面打钩，单击 OK 按钮，进入 Choose Destination Directory 界面，并选择测试类存放的位置，如图 2.4.7 和图 2.4.8 所示。

图 2.4.7　创建测试类

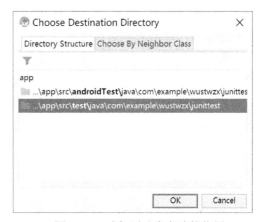

图 2.4.8　选择测试类存放的位置

（6）此时，在 junittest(test)目录下将自动生成测试类 CalculatorTest.java 文件，编写测试类 CalculatorTest.java 的测试代码（加粗文本）如下：

```
package com.example.wustwzx.junittest;
import org.junit.Test;
import static org.junit.Assert.*;
public class CalculatorTest {
    Calculator calculator = new Calculator();
    @Test
    public void add() {
        int addResult = calculator.add(2,1);      //方法返回结果
        assertEquals(3,addResult);                //断言
        //System.out.println(addResult);
    }
    @Test
    public void subtract() {
        int subResult = calculator.subtract(2,1);
        assertEquals(1,subResult);
    }
}
```

（7）右击 CalculatorTest 类，选择"Run CalculatorTest"项，对该测试类里的所有方法进行测试，结果如图 2.4.9 所示。

图 2.4.9　测试结果

其中，绿条表示 CalculatorTest 类的 add()方法与 subtract()方法断言正确（因为 2+1=3、2-1=1）。如果将被测试类 CalService.java 的 Add()方法中"+"换成"−"，则测试结果为红条，表示断言错误（因为 2-1=1≠3）。

注意：

（1）测试类可存放的位置有 AndroidTest 与 Test 两个目录，存放在 AndroidTest 目录的测试类必须将应用部署至手机后才能执行测试。

（2）也可用右击 add()方法或 subtract()方法，对其进行测试。

习 题 2

一、判断题

1. 文件夹 res 里的文件，都可以通过资源类 R 来访问。
2. 一个 Android 项目里，只能包含一个模块。
3. Android 平台使用 Java 虚拟机 JVM。
4. Android Studio 项目使用 Gradle 作为构建工具。
5. 一个 Android 项目中的所有源程序必须放在一个包里。

二、选择题

1. 快速查看 Android API，应选择的项目视图是_____。
   A．Projects　　　　B．Android　　　　C．Packages　　　　D．Tests
2. Android 项目的清单文件根标签是_____。
   A．<activity>　　　B．<application>　　C．<intent-filter>　　D．< manifest >
3. 下列选项中，不属于 Android 四大组件的是_____。
   A．Content　　　　　　　　　　　　B．Server
   C．ContentProvider　　　　　　　　D．BroadcastReceiver
4. 真正在 Android 虚拟机上运行的文件类型是_____。
   A．exe　　　　　　B．dex　　　　　　C．class　　　　　　D．odex
5. 将 Android 的运行信息显示于计算机，需要使用 Android 提供的_____工具。
   A．AVD　　　　　　B．SDK　　　　　　C．ADB　　　　　　D．ADT
6. 在 Android Studio 工具栏里，用于构建项目的工具是_____。
   A． 　　　　　　　B． 　　　　　　　C． 　　　　　　　D．

三、填空题

1. 项目清单文件里，标签<manifest>的_____属性值定义 Android 应用的包名。
2. Android 清单文件或布局里，对图片文件或字符串的引用，使用前导符号_____。
3. Android 程序运行窗口上方的标题，由清单文件里<application>标签使用_____属性而指定。
4. 在项目清单文件中注册了多个 Activity 时，只有_____Activity 含有<intent-filter>标签。
5. 在 Activity 组件里，使用其方法_____呈现布局的内容视图。

实 验 2

一、实验目的
（1）掌握 Android Studio 项目的文件系统结构及常用视图。
（2）初步掌握 Android 应用程序的基本组成部分。
（3）掌握 Android 应用程序的多种调试方法。

二、实验内容及步骤

1. 掌握 Android Studio 项目的文件系统结构

（1）在 Android Studio 中，打开上次实验创建的 HelloAndroid 项目，然后打开其清单文件。

（2）查看应用图标的设置方法，它引用了位于文件夹 res/mipmap 里的矢量图形文件 ic_launcher.xml。通过右击文件夹 res/mipmap 或 res/drawable→New→Image Asset 或 Vector Asset 的方式，更换项目的图标。

（3）查看和修改应用的标题，需要打开位于文件夹 res/values 里的键值对文件 strings.xml。

（4）打开存放在文件夹 res/layout 里.xml 格式的布局文件，选择 Text 视图并右击上方的 Preview 工具，以方便查看布局控件及其代码的对应关系。

（5）源程序 MainActivity.java 的 onCreate()方法里，在设置内容视图语句前，添加一条语句 setTitle()，重设当前 Activity 组件的标题，部署项目后做运行测试。

（6）查看项目清单文件，定义 Android 应用包名的位置。

（7）使用 File→New→Activity→Basic Activity，创建另一个 Activity 组件后，查看项目清单文件里的 Activity 注册信息，初步掌握<intent-filter>标签的作用。

（8）在当前项目里，使用 File→New→New Module，创建一个名 example2_1 的模块，编写调用系统拨号程序的应用。

（9）在工具▶图标的左边，选择当前模块为 example2_1，然后部署本模块并做运行测试。

2. 掌握 Android Studio 项目的调试方法

（1）在 HelloAndroid 项目的 onCreate()方法里，编写 Toast 信息的代码。

（2）编写只定义类成员类型但未实例化的引用代码，运行程序时将闪退，分别通过 Run 控件台和 Logcat 控制台并选择 Error 级别，查看出错信息。

（3）在程序某些代码行前设置断点，单击工具栏的 ❈ 工具，以调试模式运行应用，进行断点跟踪。

（4）根据例 2.4.1 的步骤，练习单元测试方法。

三、实验小结及思考

（由学生填写，重点写上机中遇到的问题。）

第 3 章　Android UI 与 Activity 组件

Activity 组件是装载可显示 Widget 控件的容器并实现其控制逻辑，它包含了用户界面的设计，是 Android 程序设计前的必备步骤。本章的学习要点如下：
- 掌握 Android 的视图模型及视图类的使用；
- 掌握利用 Intent 对象实现应用程序间共享数据的方法；
- 掌握 android.content.Intent 类提供的常用动作名称及数据；
- 掌握单击事件监听器的多种使用方法；
- 掌握 Toolbar、OptionMenu 菜单、Fragment、底部导航和数据适配器的使用方法。

3.1　用户界面 UI 设计

3.1.1　Android 界面视图类

Android 图形化的用户界面（Graphical User Interface，GUI）采用了结构清晰的 MVC 模型（Model-View-Controller），其具体含义：
- 提供了处理用户输入的控制器（Controller）；
- 显示用户界面的视图（View）；
- 保存数据和代码的模型（Model）。

在 Android MVC 中，控制器是由 Activity 组件完成的，它能够接受并响应程序的外部动作，如按键动作或触摸屏动作等，每个外部动作作为一个对立的事件被加入队列中，按照"先进先出"的规则从队列中获取事件，并将这个事件分配给所对应的事件处理函数。

控制器负责接受并响应程序的外部动作；通过视图反馈应用程序给用户的信息（通常是手机屏信息反馈）；模型是应用程序的核心，用于保存数据和代码。

Android 界面元素的组织形式采用视图树模型，如图 3.1.1 所示。

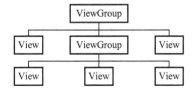

图 3.1.1　Android 界面元素的树状组织形式

Android 的视图类 android.view.View 提供了用于处理屏幕事件的多个内部接口（如 OnClickListener 等）及常用方法（如 setVisibility()等），如图 3.1.2 所示。

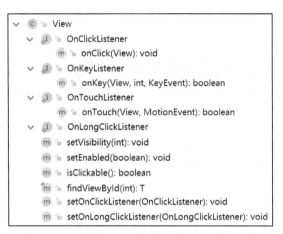

图 3.1.2　视图类 View 的内部接口及常用方法

注意：View 是一个矩形区域，它负责区域里元素的绘制；ViewGroup 是 View 的子类，它是一个容器，专门负责布局，本身没有可绘制的元素。

3.1.2　Android 用户界面事件

在 Android 系统中，各种屏幕手势的相关信息（如操作类别、发生时间等）被自动封装成一个 KeyEvent 对象，供应用程序使用。因此，在 Activity 的事件处理方法中，需要使用表示手势事件对象的事件参数 event。

Activity 提供了响应各种屏幕手势的方法，如按键方法 onKeyDown()、松开按键方法 onKeyUp()、长按键方法 onKeyLongPress()等，如图 3.1.3 所示。

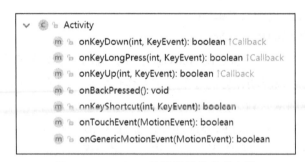

图 3.1.3　Activity 的屏幕手势处理方法

在图 3.1.3 中事件函数的返回值都是 boolean 型。当返回 True 时，表示已经完整地处理了这个事件，并不希望其他的回调方法再次进行处理；而当返回 False 时，表示没有处理完该事件，希望其他回调方法继续对其进行处理。

除了使用界面事件外，还有键盘事件。类 KeyEvent 定义了分别对应于 Back 键、Home 键和 Menu 键的键盘扫描码（实质上是静态常量），如图 3.1.4 所示。

图 3.1.4 类 KeyEvent 定义的手机键盘常量

Activity 提供的方法 onKeyDown(keyCode, event)用来捕捉手机键盘被按下（短按，不是长按）的事件。其中，参数 keyCode 表示键盘扫描码，通常使用类 KeyEvent 的静态常量表示。

默认情况下，用户按返回键将关闭一个 Activity 或返回上一级 Activity。有时，在一个 Activity 中，需要禁用返回键，这时应添加如下代码：

```
@Override
public boolean onKeyDown(int keyCode, KeyEvent event) {
    // TODO Auto-generated method stub
    if(KeyEvent.KEYCODE_BACK == keyCode)       //返回键
        return false;                          //屏蔽
    return super.onKeyDown(keyCode, event);    //其他按键默认处理
}
```

Android 程序通常需要侦听用户和应用程序之间交互的事件。对于用户界面中的事件，侦听方法就是从与用户交互的特定视图对象中截获这些事件。

事件侦听器（Event Listener）是视图 View 类的内部接口，每个内部接口包含一个单独的回调方法。这些方法将在视图中注册的侦听器被用户界面操作触发时由 Android 框架调用。下面这些回调方法被包含在事件侦听器接口中：

- onClick()包含于 View.OnClickListener 中，单击时调用；
- onLongClick()包含于 View.OnLongClickListener 中，长按时调用；
- onTouch()包含于 View.OnTouchListener 中，当用户执行的动作被当作一个触摸事件时被调用，包括按下、释放和在屏幕上进行的任何移动手势。

注意：回调方法 onClick()没有返回值，但是其他事件侦听器必须返回一个布尔值。返回 True 表示已经处理了这个事件而且到此为止；返回 False 表示还没有处理并继续交给其他事件侦听器处理。

对于一个具备触摸功能的设备，一旦用户触摸屏幕，设备将进入触摸模式。用户可以调用方法 isInTouchMode()来检测设备当前是否处于触摸模式。

Android 应用根据用户输入处理常规的焦点移动，这包含当视图删除或隐藏，或者新视图出现时改变焦点。请求一个接受焦点的特定视图，需要调用方法 requestFocus()。

3.1.3 界面与布局

res/layout 目录下存放定义 UI 设计的 XML 文件。UI 设计有两种方式：一种是 Text 视图方式，在 XML 文件中，可以直接写布局及控件代码；另一种是 Design 视图方式，可以直接拖曳控件至设计区域。Android Studio UI 设计工作界面如图 3.1.5 所示。

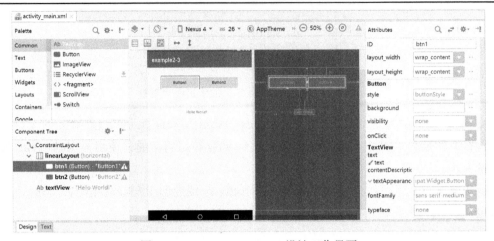

图 3.1.5　Android Studio UI 设计工作界面

注意：

（1）在实际开发中，经常切换这两种方式。纯代码方式要求用户对各种 UI 控件名称及属性非常清楚，而拖曳方式则比较快捷和直观。

（2）在 Text 视图方式下，使用 Android Studio UI 设计界面右上方的 PreView 工具，可以实现代码与设计的同步显示。

同步显示布局设计代码及设计效果的示例，如图 3.1.6 所示。

图 3.1.6　UI 设计代码与设计效果同步

布局相当于一个容器控件，其属性 android:padding（或 android:paddingLeft 等）用于控制该容器内第一个元素与父布局（容器）之间的间隔，而属性 android:layout_margin 等用于设置同一布局内各元素之间的间隔。布局内控件的常用布局属性还有以下几个。

- android:layout_width 表示控件的宽度。
- android:layout_height 表示控件的高度。

- android:id 表示控件的标识。
- android:layout_below 表示将该控件的底部置于给定 ID 控件之下。

注意：

（1）UI 设计有顶层容器和中间容器之分，根布局对应于顶层容器。

（2）控件的宽度和高度以 dp（常用于图像）或 sp（常用于文本）等为单位。

1. 线性布局

线性布局使用成对的<LinearLayout>标签，必须指定分别表示容器宽度和高度的两个属性 android:layout_width 和 android:layout_height，它们的取值为 match_parent（匹配父容器）、wrap_content（自适应控件大小）或具体值（以 dp 为单位）。

线性布局可选属性主要是 android:orientation，当取值为 vertical 时称为垂直线性布局，当取值为 horizontal 时称为水平线性布局，且以 vertical 为默认值。

垂直线性布局时，每个控件占一行；水平线性布局时，控件自左向右排列，控件太多时也不会转行。

注意：调整线性布局内的各个控件之间的间距，使用 android:layout_weight 属性比较方便。各个控件的 android:layout_weight 属性的默认值为 0，表示按照实际大小布局。

线性布局也可以嵌套使用。例如，在垂直线性布局里嵌套一个水平线性布局时，就可以在一行内水平放置多个控件。

在水平排列的 LinearLayout 里，产生控件之间的分隔线，设置如下：

```
showDividers="middle"：指定在组件之间使用分隔条
divider="?android:attr/dividerHorizontal"：设置线条
```

2. 相对布局

相对布局在布局文件中使用<RelativeLayout>标签，除了第一个元素外，其他元素需要参考另一个元素进行相对定位（含方向、偏移和对齐方式），常用属性如下。

- android:layout_below 表示位于下方。
- android:layout_above 表示位于上方。
- layout_toRightOf 表示位于右方。
- layout_toLeftOf 表示位于左方。
- android:layout_marginTop 表示偏移（正值向下，负值向上）。
- android:layout_marginLeft 表示偏移（正值向右，负值向左）。
- android:layout_alignLeft 表示左对齐（默认）。
- android:layout_alignRight 表示右对齐。
- android: layout_alignTop 表示顶部对齐（默认）。
- android:layout_ alignBottom 表示底部对齐。
- android: layout_alignBaseline 表示垂直居中。

注意：

（1）如果相对布局内的多个元素没有进行相对定位，则会重叠（只显示一个）。

（2）当布局内元素较多或布局不规则时，使用相对布局较方便且不需要嵌套。

（3）对齐可以通过偏移来实现。

（4）应用属性 android:layout_margin，相当于同时应用了 android:layout_marginTop 和 android:layout_marginLeft 两个属性。

3．约束布局

传统的 Android 开发，其界面基本都是靠编写 XML 代码完成的。约束布局是 Android Studio 2.3 之后推荐且默认使用的布局，完全使用可视化的操作方式，在布局文件中使用 <android.support.constraint.ConstraintLayout>标签声明约束布局。

当拖曳某个控件至界面时，系统提示缺少约束的红色警告，表明程序运行时该控件将出现在屏幕左上角(0,0)位置。解决方法有如下两种：

（1）单击该控件，使用工具 来自动添加用于确定控件位置的约束线。此时，该控件至少存在两条约束线；

（2）单击该控件，从出现的四个圆点中，任选一个圆点并拖曳到四周的边线或控件圆点来建立位置约束。重复此操作，直到红色警告消失。

选中某个已经添加约束的控件，若使用弹出的工具 ，将删除已经建立的所有约束线；若单击其中的某条约束线的圆点端，将删除本约束线。

注意：

（1）在 res/layout 里新建布局文件时，默认使用的根标签是 LinearLayout。如果要使用约束布局，则需要更换根标签为 ConstraintLayout。

（2）线性布局可作为一个容器控件并添加至约束布局里，也要进行位置约束。

4．帧布局

帧布局像一层层画布，添加的控件一层层地放上去。帧布局添加的各个控件默认都将对齐到屏幕的左上角。

在一个有两层的帧布局中，如果浮于上面的第二层可以看到下层，是因为 background 属性设置包含了透明度参数，有如下两种实现方式。

方法一是设置控件的背景色，使用如下属性：

```
android:background="#aarrggbb"
```

其中，aag 表示透明度参数（十六进制）。

方法二是在程序里使用如下方法：

```
v.getBackground().setAlpha(a);      //透明度参数 a 取值范围为 0～255
```

其中，v 为 View 对象，a 为十进制数。

【例 3.1.1】 Android Studio UI 设计示例。

一个使用约束布局并内嵌线性布局，包含文本框、复选框、下拉列表和命令按钮的组件树及其设计视图，如图 3.1.7 所示。

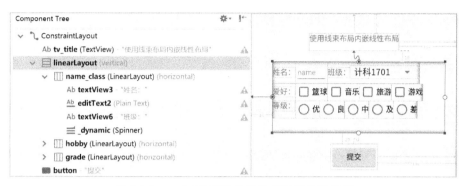

图 3.1.7　一个 UI 设计的组件树及其设计视图

先建立一个名为 array.xml 的 XML 文件，并存放至 res/values 文件夹里，文件代码如下：

```xml
<?xml version="1.0" encoding="utf-8"?>
<resources>
    <!--定义一个名为 class_array 的字符串数组-->
    <string-array name="class_array">
        <item>计科 1701</item>
        <item>计科 1702</item>
        <item>软件 1701</item>
        <item>网络 1701</item>
    </string-array>
</resources>
```

相应的布局代码如下：

```xml
<?xml version="1.0" encoding="utf-8"?>
<android.support.constraint.ConstraintLayout
xmlns:android="http://schemas.android.com/apk/ res/android"
    xmlns:app="http://schemas.android.com/apk/res-auto"
    xmlns:tools="http://schemas.android.com/tools"
    android:layout_width="match_parent"
    android:layout_height="match_parent"
    tools:context=".MainActivity"
    tools:layout_editor_absoluteY="81dp">
    <TextView
        android:id="@+id/tv_title"
        android:layout_width="wrap_content"
        android:layout_height="32dp"
        android:layout_marginEnd="108dp"
        android:layout_marginTop="36dp"
        android:text="使用线束布局内嵌线性布局"
        app:layout_constraintEnd_toEndOf="parent"
        app:layout_constraintTop_toTopOf="parent" />
```

```xml
<LinearLayout
    android:id="@+id/linearLayout"
    android:layout_width="323dp"
    android:layout_height="131dp"
    android:layout_marginStart="44dp"
    android:layout_marginTop="60dp"
    android:orientation="vertical"
    app:layout_constraintStart_toStartOf="parent"
    app:layout_constraintTop_toBottomOf="@+id/tv_title">
    <LinearLayout
        android:id="@+id/name_class"
        android:layout_width="wrap_content"
        android:layout_height="wrap_content"
        android:orientation="horizontal">
        <TextView
            android:id="@+id/tv_name"
            android:layout_width="wrap_content"
            android:layout_height="wrap_content"
            android:layout_weight="1"
            android:text="姓名："
            android:textSize="14sp" />
        <EditText
            android:id="@+id/et_name"
            android:layout_width="60dp"
            android:layout_height="wrap_content"
            android:layout_weight="1"
            android:ems="10"
            android:hint="name"
            android:inputType="textPersonName"
            android:textSize="14sp" />
        <TextView
            android:id="@+id/tv_class"
            android:layout_width="wrap_content"
            android:layout_height="wrap_content"
            android:layout_weight="1"
            android:text="班级：" />
        <!--需要先建立一个名为 class_array 的字符串数组，并存在 values 里的.xml 文件里-->
        <Spinner
            android:id="@+id/dropdown_class"
            android:layout_width="wrap_content"
            android:layout_height="wrap_content"
            android:entries="@array/class_array"/>
    </LinearLayout>
    <LinearLayout
        android:id="@+id/hobby"
        android:layout_width="wrap_content"
```

第3章 Android UI 与 Activity 组件

```
        android:layout_height="wrap_content"
        android:orientation="horizontal">
    <TextView
        android:id="@+id/tv_hobby"
        android:layout_width="wrap_content"
        android:layout_height="wrap_content"
        android:layout_weight="1"
        android:text="爱好：" />
    <CheckBox
        android:id="@+id/checkBox"
        android:layout_width="wrap_content"
        android:layout_height="wrap_content"
        android:layout_weight="1"
        android:text="篮球" />
    <CheckBox
        android:id="@+id/checkBox2"
        android:layout_width="wrap_content"
        android:layout_height="wrap_content"
        android:layout_weight="1"
        android:text="音乐" />
    <CheckBox
        android:id="@+id/checkBox3"
        android:layout_width="wrap_content"
        android:layout_height="wrap_content"
        android:layout_weight="1"
        android:text="旅游" />
    <CheckBox
        android:id="@+id/checkBox4"
        android:layout_width="wrap_content"
        android:layout_height="wrap_content"
        android:layout_weight="1"
        android:text="游戏" />
</LinearLayout>
<LinearLayout
    android:id="@+id/grade"
    android:layout_width="wrap_content"
    android:layout_height="wrap_content"
    android:orientation="horizontal">
    <TextView
        android:id="@+id/tv_grade"
        android:layout_width="wrap_content"
        android:layout_height="wrap_content"
        android:text="等级：" />
    <RadioGroup
        android:id="@+id/grade_group"
        android:layout_width="wrap_content"
        android:layout_height="wrap_content"
```

```xml
            android:layout_weight="0"
            android:orientation="horizontal">
            <RadioButton
                android:id="@+id/radioButton"
                android:layout_width="wrap_content"
                android:layout_height="wrap_content"
                android:text="优" />
            <RadioButton
                android:id="@+id/radioButton2"
                android:layout_width="wrap_content"
                android:layout_height="wrap_content"
                android:layout_weight="1"
                android:text="良" />
            <RadioButton
                android:id="@+id/radioButton3"
                android:layout_width="wrap_content"
                android:layout_height="wrap_content"
                android:layout_weight="1"
                android:text="中" />
            <RadioButton
                android:id="@+id/radioButton4"
                android:layout_width="wrap_content"
                android:layout_height="wrap_content"
                android:layout_weight="1"
                android:text="及" />
            <RadioButton
                android:id="@+id/radioButton5"
                android:layout_width="wrap_content"
                android:layout_height="wrap_content"
                android:layout_weight="1"
                android:text="差" />
        </RadioGroup>
    </LinearLayout>
</LinearLayout>
<Button
    android:id="@+id/button"
    android:layout_width="wrap_content"
    android:layout_height="wrap_content"
    android:layout_marginEnd="148dp"
    android:layout_marginTop="40dp"
    android:text="提交"
    app:layout_constraintEnd_toEndOf="parent"
    app:layout_constraintTop_toBottomOf="@+id/linearLayout" />
</android.support.constraint.ConstraintLayout>
```

注意：

（1）本布局文件\<Spinner\>标签的 entries 属性，使用了名为 class_array 的字符串数组，指定下拉列表的数据源（列表项）。

（2）Spinner 控件的数据源也可以在 Activity 程序里通过应用数据适配器的方式指定。

（3）Spinner 控件默认使用 dropdown 模式来展示列表项。实际上，Spinner 控件还可以使用 dialog 模式。

3.2 活动组件 Activity

3.2.1 AppCompatActivity、Activity 和 Context

1. AppCompatActivity

随着 Android 开发技术的发展，在不同的阶段（开发环境）创建 Activity 组件所使用的基类不同。在使用 eclipse 进行 Android 开发时，自动创建的 MainActivity 继承 Activity，而 Android Studio 继承 AppCompatActivity。

注意：

（1）Android 6.0 推出之后，提供了很多新东西，于是 Support v7 也更新了，出现了 AppCompatActivity，它是 Activity 的子类。

（2）Android 6.0 动态权限处理，需要使用 AppCompatActivity。

Activity 作为 Android 最重要的组件之一，用于设计应用程序的用户界面，其内容来源于布局文件。在一个 Activity 的 onCreate()方法里，使用父类的方法 setContentView()呈现内容视图，并以布局文件参数。Activity 包含了响应界面事件的代码，即具有控制器的功能。

在 Android Studio 编辑窗口中，选中 AppCompatActivity 并按【Ctrl+H】组合键时，将显示类 AppCompatActivity 的继承关系，如图 3.2.1 所示。

图 3.2.1　类 AppCompatActivity 的超类

注意：在使用 Android 的四大组件前，必须在项目清单文件中注册相应的权限。

2. Activity 作为上下文类的子类

在 Java SE 中，创建一个类，写个 main()方法就能运行。一个类调用另一个类，需要先使用 new 运算符创建另一个类的实例对象。Android 应用使用基于组件的设计模式，组件的运行要有一个完整的 Android 项目环境，在这个环境下，Activity 和 Service 等系统组件才能够正常工作，而这些组件并不能采用普通的 Java 对象创建方式（使用 new 运行创建实例），而是要有它们各自的上下文环境。Android 提供了维持程序中各组件能够正常工作的一个核心功能类——上下文类 android.content.Context。

抽象类 Context 是 Activity 的父类，但不是直接父类，因为 Java 只支持单继承。因此，Activity 继承（具有）Context 类的所有方法。Context 类具有的主要方法如图 3.2.2 所示。

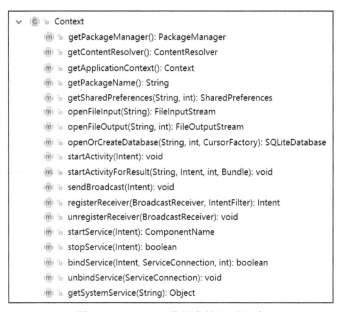

图 3.2.2　Context 类具有的主要方法

3.2.2　Activity 组件的基本方法

Activity 组件作为用户界面，其子类 AppCompatActivity 提供的基本方法如图 3.2.3 所示。

图 3.2.3　AppCompatActivity 提供的基本方法

3.2.3　Activity 类具有的扩展方法

在 Android 应用开发中，除了必须掌握 Activity 具有的基本方法外，还需要掌握一些扩展的方法，如用于数据共享存储的 getPreferences()方法、用于动态加载页面布局的

getLayoutInflater()方法、开始后台服务的 startService()方法、发送广播的 sendBroadcast() 方法、获得内容解析器方法 getContentResolver()等，如表 3.2.1 所示。

表 3.2.1　Activity 类具有的扩展方法

方　法　名	功　能　描　述
getPreferences() getSharedPreferences()	得到 SharedPreferences 对象，使用 XML 文件存放数据，文件存放在/data/data/<package name>/shared_prefs 目录下
getLayoutInflater()	Activity 类的方法，将布局文件实例化为 View 类对象，实现动态加载布局
getSystemService()	根据传入的系统服务名称得到相应的服务对象
startActivity()	Activity 调用
startActivityForResult()	有返回值 Activity 调用
startService()/bindService()	继承的方法，启动某个服务
sendBroadcast()/unregisterReceiver()	继承的方法，发送广播/取消注册的广播接收者
getContentResolver()	继承的方法，得到一个 ContentResolver 对象
getPackageManager()	继承的方法，得到一个管理和查询系统所有应用程序的 PackageManager 对象

注意：在上面的方法中，除了涉及 Activity 组件外，还涉及 Android 的另外三个组件：服务组件（Service）、广播接收者组件（BroadcastReceiver）和内容提供者组件（ContentProvider）。

布局膨胀器类 LayoutInflater 提供了 inflate()方法，用在 Activity 组件程序里，实现视图的动态加载，其主要代码如下：

```
LayoutInflater inflater = this.getLayoutInflater();   //获取 LayoutInflater 对象
View layoutView = inflater.inflate(R.layout.login_view, null); //加载布局文件
//将 layoutView 视图加载到主视图
```

3.2.4　Activity 的生命周期

复杂的 Android 应用可能包含若干个 Activity。当打开一个新的 Activity 时，先前的那个 Activity 会被置于暂停状态，并压入历史堆栈中。用户可以通过返回键退到先前的 Activity。

Activity 是由 Android 系统维护的，每个 Activity 除了有创建 onCreate()、销毁 onDestroy()两个基本方法外，还有激活方法 onStart()、恢复方法 onResume()、暂停方法 onPause()、停止方法 onStop()和重启方法 onRestart()。

Activity 在其生命周期中存在三种状态：运行态、暂停态和停止态。运行态是指 Activity 调用 onStart()方法后出现在屏幕最上层的状态，此时用户通常可以获取焦点；暂停态是指 Activity 调用 onPause()方法之后出现的状态，其上还有处于运行态的 Activity 存在，并且 Activity 没有被完全遮挡住，即处于暂停态的 Activity 有一部分视图被用户所见；停止态是指当前 Activity 调用 onStop()之后的状态，此时它完全被处于运行态的

Activity 遮挡住，即程序界面完全不被用户所见。

对于处于运行态的 Activity，当用户按返回键退出时，将依次调用生命周期方法 onStop()和 onDestroy()来销毁当前的 Activity。

Activity 组件的生命周期如图 3.2.4 所示。

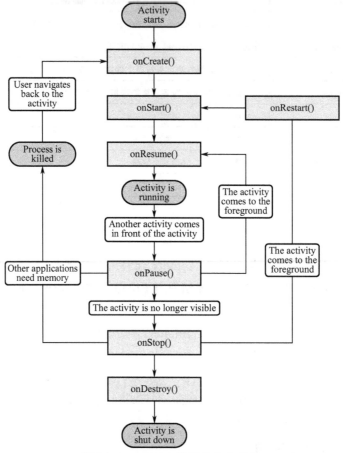

图 3.2.4　Activity 组件的生命周期

注意：

（1）处于暂停态或停止态的 Activity 在系统资源缺乏时可能被杀死，以释放其占用的资源。

（2）在 Android Studio 中编辑 Activity 组件时，按【Ctrl+O】组合键可选择需要重写的父类方法。

3.2.5　手机横/竖屏自动切换问题

目前的手机默认设置能进行横/竖屏自动切换。例如，在手机上播放视频时，通常使用横屏。此时，只需要旋转一下手机即可。

如果要关闭横/竖屏自动切换功能，有以下几种选择方式。

（1）运行手机设置程序，取消竖屏自动切换功能。此时，所有应用都将禁止竖屏自动切换，除非程序里打开了竖屏自动切换功能。

（2）在应用的主 Activity 程序的 onCreate()方法里，使用如下代码设置禁止横屏：

setRequestedOrientation(ActivityInfo. SCREEN_ORIENTATION_PORTRAI);   //始终竖屏

（3）在清单文件里注册主 Activity 时，使用如下属性来强制竖屏：

android:screenOrientation="portrait"   <!--强制竖屏-->

3.3 常用 Widget 控件的使用

android.widget 包内提供了大量用于 UI 设计的控件。实际上，在创建 Android 项目时布局文件中就含有一个用于显示文本的文本框控件 TextView。下面分别介绍这些 Widget 控件的使用方法。

在 Android Studio 控件箱里，对控件进行了分类，分为 Common、Text 和 Buttons 等类，如图 3.3.1 所示。

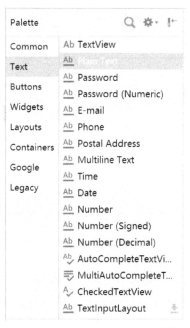

图 3.3.1　Android Studio 控件箱

3.3.1　文本控件 TextView 和 EditText

常用的 Text 类控件是 TextView 和 EditText。TextView 是用来显示字符的控件，而 EditText 是用来输入和编辑字符的控件。

注意：控件工具 Ab TextView 产生的控件标签是<TextView>，而 Ab Plain Text 产生的标签是<EditText>。

类 TextView 主要提供了 setText()方法，如图 3.3.2 所示。

图 3.3.2　TextView 类的主要方法

注意：使用方法 setText(v)赋值时，如果 v 的类型为 int 等类型，则由于参数类型一致而导致运行时异常（程序闪退），尽管没有语法错误和编译错误提示。此时，解决的办法是使用 v+""代替 v。

在布局文件中，可以使用文本框控件的常用属性，如表 3.3.1 所示。

表 3.3.1　文本框控件的常用属性

方 法 名	功 能 描 述
text 属性	文本框内的文本
layout_width 属性	文本框的宽度
layout_height 属性	文本框的高度
password 属性	属性为"true"时，表示使用密码输入形式（已经不推荐使用）
singleLine 属性	属性为"false"时，表示多行文本

使用 EditText 控件的 hint 属性，实现输入前的文本提示并在输入任意字符后消失。

当 TextView 的内容特别多时，可使用它的一个重要方法 setMovementMethod(new ScrollingMovementMethod())实现滑动，即可以通过手指的上下滑动来查看内容。当在布局文件中使用属性 android:scrollbars="vertical"时，就会在文本框内的边上产生起提示作用的垂直滑块。

注意：

（1）EditView 是 TextView 的子类，EditText 是一个具有编辑功能的 TextView 控件。

（2）setMovementMethod(new ScrollingMovementMethod())实现的滑动效果，也可以通过使用容器控件 ScrollView 来实现。

（3）文本框内文字的大小，一般使用 sp 作为单位。

（4）当 ListView 控件内容超过一屏的容量时，会自动产生滚动条。

3.3.2　图像控件 ImageView

控件 ImageView 用来显示图像。在布局文件中，使用标签<ImageView>属性 android:src 或 android:srcCompat 指定图像的来源，也可在程序里使用方法 setImageResource(resId)动态地指定图像的来源。

此外，ImageView 控件有宽度和高度等属性。

注意：ImageView 控件的图像来源也可以是手机外部存储或网络。此时，会涉及权限或专门的图像框架（如 Glide，参见 11.2.1 节）。

3.3.3 命令按钮控件 Button、ImageButton 及其单击事件监听器设计

Button 是 UI 设计中使用相当频繁的一个控件，用来定义命令按钮。本来，任何 View 对象都可以设置单击事件监听器，只是 Button 控件有特定的外观，并且文本默认居中。

注意：按钮的英文字母默认全部大写，必要时对 Button 控件应用如下属性：

```
android:textAllCaps="false"
```

当用户单击按钮时，会有相应的动作，其动作代码放在按钮的单击事件监听器的 onClick()方法内。Button 控件对象通过 setOnClickListener()方法设置单击事件监听器，方法参数为一个实现了 android.view.View.OnClickListener 接口的对象。

单击事件监听器有多种使用方式，当 Activity 组件只有一个 Button 按钮时，一般是在 setOnClickListener()方法里创建一个匿名内部类对象，其示例代码（写在组件的 onCreate()方法里）如下：

```java
//下面的 Button 为布局文件里 Button 控件的 ID
findViewById(R.id.button).setOnClickListener(new View.OnClickListener() {
    @Override
    public void onClick(View v) {
        //单击按钮后的业务逻辑
    }
});
//下面是等效代码：先创建接口类型的对象，后应用于监听器方法
View.OnClickListener listener = new View.OnClickListener() {
    @Override
    public void onClick(View v) {
        //单击按钮后的业务逻辑
    }
};
findViewById(R.id.button).setOnClickListener(listener);
```

当一个 Activity 中需要定义单击事件的对象较多时，通常在定义 Activity 组件时，通过子句 implements View.OnClickListener 定义 Activity 组件也是该接口的实现类，其示例代码如下：

```java
public class MainActivity extends AppCompatActivity implements View.OnClickListener{
    @Override
    void onCreate(Bundle savedInstanceState) {
        super.onCreate(savedInstanceState);
        setContentView(R.layout.activity_main);
        //先找到 Button 控件对象，然后设置监听（this 代表当前对象）
        Button btn1 = findViewById(R.id.btn1);    btn1.setOnClickListener(this);
        Button btn2 = findViewById(R.id.btn2);    btn2.setOnClickListener(this);
        Button btn3 = findViewById(R.id.btn3);    btn3.setOnClickListener(this);
    }
    @Override
```

```java
        public void onClick(View view) {        //集中处理监听
            switch (view.getId()) {
                case R.id.btn1:
                    //按钮 btn1 的业务逻辑
                    break;
                case R.id.btn2:
                    //按钮 btn2 的业务逻辑
                    break;
                case R.id.btn3:
                    //按钮 btn3 的业务逻辑
            }
        }
    }
```

除了上面两种用法外,还可以写 View.OnClickListener 接口的实现类作为 Activity 组件的内部类并作为其成员,再创建该内部类的实例作为监听器方法的参数。示例代码如下:

```java
@Override
protected void onCreate(Bundle savedInstanceState) {
    super.onCreate(savedInstanceState);
    setContentView(R.layout.activity_main);
    Button btn = findViewById(R.id.btn);
    btn.setOnClickListener(new MyListener());
}
class MyListener implements View.OnClickListener{
    @Override
    public void onClick(View v){
        //单击事件后执行的业务逻辑
    }
}
```

注意:

(1) OnClickListener 是类 android.view.View 的内部接口。

(2) 对 View 对象应用 setOnClickListener(null),相当于没有监听 View 对象。

(3) 对 TextView 控件对象(显示文本)或 ImageView 控件对象(显示图像)也能设置单击事件监听器,尤其是文本的单击事件监听器在手机 App 中大量使用。

3.3.4　单选按钮控件 RadioButton 与复选框控件 CheckBox

RadioGroup 为单选按钮组控件,它包含若干单选按钮控件 RadioButton。通过这两个控件,可以为用户提供一种"多选一"的选择方式。

CheckBox 为复选框控件,通过使用父类 CompoundButton 提供的方法 isChecked() 来判断用户是否勾选该复选框。

RadioGroup、RadioButton 和 CheckBox 的定义如图 3.3.3 所示。

第 3 章 Android UI 与 Activity 组件

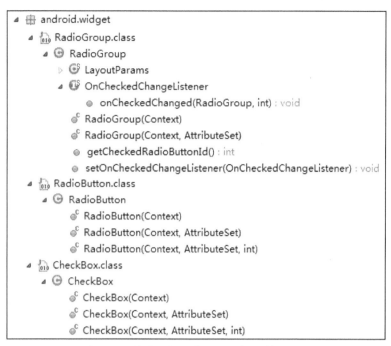

图 3.3.3　单选按钮控件与复选框控件的定义

注意：

（1）使用单选按钮的布局文件中，一个 RadioGroup 控件内嵌若干 RadioButton 控件。

（2）RadioButton 和 CheckBox 控件都有说明选项含义的 android:text 属性。

（3）RadioButton 和 CheckBox 都继承了抽象类 CompoundButton。CompoundButton 类的定义如图 3.3.4 所示。

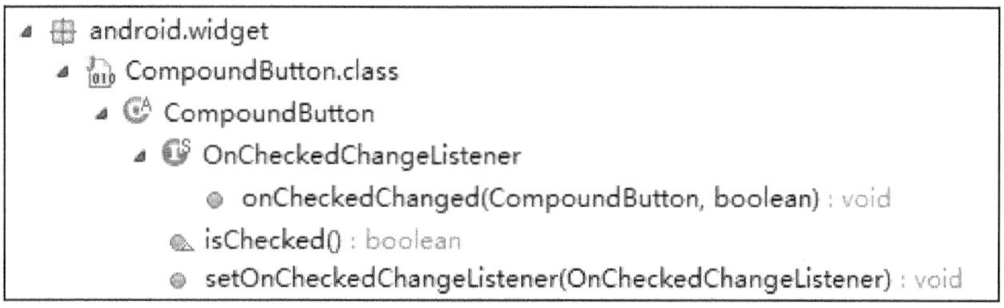

图 3.3.4　CompoundButton 类的定义

3.3.5　消息提醒对话框控件 AlertDialog 与进度控件 ProgressDialog

对话框与 Toast 一样，也属于消息提醒，但它是以对话框的形式出现在屏幕上的，用于需及时处理的通知。

对话框控件是特殊的控件，且有多种表现形式，它们不同于 TextView 等控件，在对话完成后才消失（需要用户干预）。因此，对话框控件 AlertDialog 并不位于 android.widget 包内，而是位于 android.app 包内。

55

注意：

（1）AlertDialog 是 Dialog 的子类，ProgressDialog 是 AlertDialog 的子类。

（2）android.widget.ProgressBar 也是进度处理类。

【例 3.3.1】 在对话框里动态加载布局视图。

程序运行时，单击命令按钮后，效果如图 3.3.5 所示。

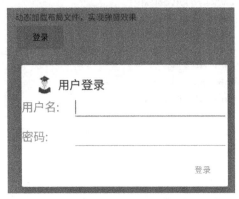

图 3.3.5　在对话框里动态加载布局示例效果

主要设计步骤如下：

（1）创建布局文件 res/layout/login_view.xml，它包含两个 EditText 控件。

（2）在主布局文件里添加一个名为 btn 的 Button 控件。

（3）在 Activity 组件里，创建一个单击监听器对象 listener 作为类成员，其重写的方法 onClick() 包含了加载布局和创建对话框；在组件的 onCreate() 方法里，对按钮 btn 应用单击监听器 listener。

Activity 组件程序的代码如下：

```java
public class MainActivity extends AppCompatActivity {
    //定义监听器对象作为类成员
    View.OnClickListener listener = new View.OnClickListener() {
        @Override
        public void onClick(View arg0) {
            LayoutInflater inflater = MainActivity.this.getLayoutInflater();//获取 LayoutInflater 对象
            View loginView = inflater.inflate(R.layout.login_view, null); //加载布局文件
            //创建对话框构造器对象
            AlertDialog.Builder builder = new AlertDialog.Builder(MainActivity.this);
            //设置对话框
            builder.setIcon(R.drawable.tb02)
                .setTitle("用户登录")
                .setView(loginView)           //在对话框里动态加载布局视图
                .setPositiveButton("登录", new DialogInterface.OnClickListener() {
                    @Override
                    public void onClick(DialogInterface arg0, int arg1) {
                        //这儿写单击"登录"按钮后的业务逻辑
```

```
                }
            });
            //显示对话框
            builder.create().show();              //create()可省略
        }
    };
    @Override
    protected void onCreate(Bundle savedInstanceState) {
        super.onCreate(savedInstanceState);
        setContentView(R.layout.activity_main);
        Button btn = findViewById(R.id.btn);
        btn.setOnClickListener(listener);
    }
}
```

注意：登录布局文件里并不包含 Button 控件，"登录"按钮是由对话框产生的。

3.3.6 列表控件及其数据适配器和列表项选择监听器

1. 列表控件 ListView 及其数据适配器

ListView 是列表数据显示控件，以列表形式显示。Android 软件包 android.widget，在提供类 ListView 的同时，还提供了用于显示列表数据的数据适配器接口 ListAdapter 及其相关类，如图 3.3.6 所示。

图 3.3.6 列表控件及其数据适配器接口（类）

注意：列表控件使用广泛但较复杂，因为它的数据项会涉及 Java 泛型。

其中，抽象类 BaseAdapter 实现了 ListAdapter 接口，同时又是类 ArrayAdapter 和 SimpleAdapter 的超类。定义一个继承 BaseAdapter 类的子类 MyAdapter 时，需要重写的方法如下：

```
public class MyAdapter extends BaseAdapter {
    @Override
    public int getCount() {
        // TODO Auto-generated method stub
        return 0;
```

```
    }
    @Override
    public Object getItem(int arg0) {
        // TODO Auto-generated method stub
        return null;
    }
    @Override
    public long getItemId(int arg0) {
        // TODO Auto-generated method stub
        return 0;
    }
    @Override
    public View getView(int arg0, View arg1, ViewGroup arg2) {
        // TODO Auto-generated method stub
        return null;
    }
}
```

ListView 控件的数据适配器都继承抽象类 BaseAdapter，可根据需要选用。例如，单列数据的显示可采用 ArrayAdapter，该类的定义如图 3.3.7 所示。

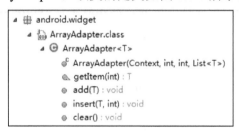

图 3.3.7　ArrayAdapter 类的定义

通常使用的构造方法是 ArrayAdapter(Context,int,int,List<T>)，其中第 1 个参数是上下文，第 2 个参数是列表控件的列表项布局文件，第 3 个参数是列表布局文件中显示数据的控件，第 4 个参数是列表数据。

如果列表项的数据项有多个，则可采用 SimpleAdapter。

注意：RecyclerView 是谷歌于 2014 年在 V7 包推出的新控件，它标准化了 ViewHolder 类，通过设置 LayoutManager 来快速实现 ListView、GridView 和瀑布流效果，还可以设置横向和纵向显示。添加动画效果也非常简单（自带 ItemAnimation），可以设置加载和移除时的动画，方便做出各种动态效果。

2．列表项选择监听器

在 Android 系统类库中，抽象类 AdapterView 是 ListView 的超类（但不是直接父类），提供了设置列表项单击事件监听器的重要方法 setOnItemClickListener()和内部接口 OnItemClickListener，用以定义用户选择了某个列表项后的动作。抽象类 AdapterView 的定义如图 3.3.8 所示。

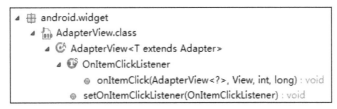

图 3.3.8 抽象类 AdapterView 的定义

【例 3.3.2】 ListView 控件及数据适配器（含列表项选择监听器）的使用。

项目主布局包含 3 个 Button 按钮和 1 个 ListView 控件，分别使用 3 种不同的数据适配器来显示列表数据。其中，使用 BaseAdapter 的运行效果如图 3.3.9 所示。

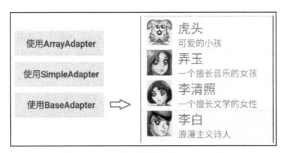

图 3.3.9 界面程序的运行效果

相应界面程序的主要代码如下：

```java
/*
    ListView 控件及其数据适配器和列表项监听
    ListView 的 setAdapter()方法的参数为 ListAdapter 接口类型
    BaseAdapter 是实现 ListAdapter 接口的抽象类
    ArrayAdapter 和 SimpleAdapter 都是 BaseAdapter 的子类
*/
public class MainActivity extends AppCompatActivity implements View.OnClickListener{
    ListView listView;
    Button button1,button2,button3;
    @Override
    protected void onCreate(Bundle savedInstanceState) {
        super.onCreate(savedInstanceState);
        setContentView(R.layout.activity_main);
        listView=findViewById(R.id.listView);
        button1=findViewById(R.id.button1);button1.setOnClickListener(this);
        button2=findViewById(R.id.button2);button2.setOnClickListener(this);
        button3=findViewById(R.id.button3);button3.setOnClickListener(this);
    }
    @Override
    public void onClick(View v) {
        int se=v.getId();
        switch (se){
```

```java
case R.id.button1:
    List<String> data1=new ArrayList<String>();
    data1.add("北京");data1.add("上海");data1.add("武汉");
    final ArrayAdapter<String> adapter1=new ArrayAdapter<String>(
            MainActivity.this,
            R.layout.item_layout1,
            R.id.tv,
            data1);
    listView.setAdapter(adapter1);
    //列表项监听
    listView.setOnItemClickListener(new AdapterView.OnItemClickListener() {
        @Override
        public void onItemClick(AdapterView<?> parent, View view,
                                int position, long id) {
            String str="你选择了"+adapter1.getItem(position);
            Toast.makeText(MainActivity.this, str, Toast.LENGTH_SHORT). show();
        }
    });
    break;
case R.id.button2:
    //列表所有数据
    List<Map<String, Object>> data2 = new ArrayList<Map<String, Object>>();
    Map<String, Object> listItem1 = new HashMap<String, Object>();
    listItem1.put("id", 1);
    listItem1.put("name", "张三");
    listItem1.put("salary", 5000);
    data2.add(listItem1);
    Map<String, Object> listItem2 = new HashMap<String, Object>();
    listItem2.put("id", 2);
    listItem2.put("name", "李四");
    listItem2.put("salary", 5800);
    data2.add(listItem2);
    Map<String, Object> listItem3 = new HashMap<String, Object>();
    listItem3.put("id", 3);
    listItem3.put("name", "王五");
    listItem3.put("salary", 5500);
    data2.add(listItem3);
    //定义数据适配器
    final SimpleAdapter adapter2 = new SimpleAdapter(
            MainActivity.this,
            data2,
            R.layout.item_layout2, //列表项布局
            new String[]{"id", "name", "salary"},          //键名数组
            new int[]{R.id.tv1, R.id.tv2, R.id.tv3});      //控件数组
    listView.setAdapter(adapter2);   //数据绑定
    listView.setOnItemClickListener(new AdapterView.OnItemClickListener() {
```

```java
            @Override
            public void onItemClick(AdapterView<?> parent, View view,
                                                     int position, long id) {
                //从数据适配器读取记录
                Map<String,Object> rec= (Map<String, Object>) adapter2.
                                                     getItem(position);
                String result=" 【"+position+"】 "+rec.get("id")+","+
                                             rec.get("name")+","+rec.get("salary");
                Toast.makeText(MainActivity.this, result,
                                             Toast.LENGTH_SHORT).show();
            }
        });
        break;
    case R.id.button3:
        String[] names = new String[]{ "虎头", "弄玉", "李清照", "李白"};
        String[] descs = new String[]{ "可爱的小孩", "一个擅长音乐的女孩",
                                       "一个擅长文学的女性", "浪漫主义诗人"};
        int[] imageIds = new int[]{ R.drawable.tiger , R.drawable.nongyu,
                                    R.drawable.qingzhao , R.drawable.libai};
        final List<Map<String, Object>> data3 = new ArrayList<Map<String, Object>>();
        for (int i = 0; i < names.length; i++) {
            Map<String, Object> listItem = new HashMap<String, Object>();
            listItem.put("header", imageIds[i]);
            listItem.put("personName", names[i]);
            listItem.put("desc", descs[i]);
            data3.add(listItem);
        }
        final BaseAdapter adapter3=new BaseAdapter() {          //BaseAdapter是抽象类
            @Override
            public int getCount() {
                return data3.size();    //记录条数
            }
            @Override
            public Object getItem(int position) {
                return data3.get(position);    //记录
            }
            @Override
            public long getItemId(int position) {
                return position;
            }
            @Override
            public View getView(int position, View convertView, ViewGroup parent) {
                //根据布局得到列表视图
                View item=View.inflate(
                                getApplicationContext(),R.layout.item_layout3,null);
                //获取列表记录的控件
```

```
                ImageView header=item.findViewById(R.id.header);
                TextView name=item.findViewById(R.id.name);
                TextView desc=item.findViewById(R.id.desc);
                Map<String, Object> p=data3.get(position);
                //按键名取值后，给控件赋值
                header.setImageResource((int)p.get("header")); //图像资源
                name.setText(p.get("personName")+"");
                desc.setText(p.get("desc")+"");
                return item;
            }
        };
        listView.setAdapter(adapter3);
        listView.setOnItemClickListener(new AdapterView.OnItemClickListener() {
            @Override
            public void onItemClick(AdapterView<?> parent, View view,
                                                        int position, long id) {
                //从数据适配器读取记录
                Map<String,Object> p= (Map<String, Object>) adapter3.
                                                              getItem(position);
                String result="你选择了："+p.get("personName");
                Toast.makeText(MainActivity.this, result,Toast.LENGTH_SHORT).show();
            }
        });
    }
}
```

3.3.7 下拉列表控件 Spinner

Spinner 控件提供下拉列表式的输入方式，可有效地节省手机屏幕的显示空间。

与 ListView 类似，Spinner 控件也有设置数据适配器方法 setAdapter()、监听选择项方法 setOnItemSelectedListener()。

Spinner 比 ListView 多一个使用方法是 setOnTouchListener()，用以实现触屏处理。当单击 Spinner 对象时才出现列表。

图 3.3.10 界面程序运行效果

注意：在 Android Studio 中，下拉列表对应于 Containers 类的工具 Spinner 产生的控件标签<Spinner>。HTML 使用<select>标签制作下拉列表，ASP.NET 使用控件<asp:DropDownList>制作下拉列表。

【例 3.3.3】 Spinner 控件的使用。

一个使用约束布局并内嵌线性布局，包含文本框、复选框、下拉列表和命令按钮的界面程序运行效果，如图 3.3.10 所示。

相应界面程序的主要代码如下：

```
findViewById(R.id.button).setOnClickListener(new View.OnClickListener() {
    @Override
    public void onClick(View v) {
        String sr = "你输入（选择）的结果是：";
        EditText et_name = findViewById(R.id.et_name);
        sr += et_name.getText() + "，";
        //下拉列表有唯一值，所有可能的取值存放在一个.xml 文件里
        Spinner sp = findViewById(R.id.dropdown_class);
        sr += sp.getSelectedItem().toString() + "，";
        //复选的值有不确定性，需要逐个判断
        CheckBox cb1 = findViewById(R.id.checkBox);
        if (cb1.isChecked()) sr += cb1.getText() + "，";
        CheckBox cb2 = findViewById(R.id.checkBox2);
        if (cb2.isChecked()) sr += cb2.getText() + "，";
        CheckBox cb3 = findViewById(R.id.checkBox3);
        if (cb3.isChecked()) sr += cb3.getText() + "，";
        CheckBox cb4 = findViewById(R.id.checkBox4);
        if (cb4.isChecked()) sr += cb4.getText() + "，";
        //单选按钮组只能选择一个值
        RadioGroup rg = findViewById(R.id.grade_group);   //先找到组
        //再从组里找选择的按钮 id
        RadioButton rb = findViewById(rg.getCheckedRadioButtonId());
        sr += rb.getText();
        //输出选择（输入）结果
        Toast.makeText(getApplicationContext(), sr, Toast.LENGTH_LONG).show();
    }
});
```

3.4 高级 UI 程序设计

3.4.1 日期和时间选择器（DatePicker 和 TimePicker）

Android 软件包 android.widget 提供了两个实用控件 DatePicker 和 TimePicker，以方便用户以图形化的方式输入日期和时间。其中，日期信息（年、月、日）通过 DatePicker 提供的 init() 方法获取；时间信息（时、分）通过 TimePicker 的内部接口 OnTimeChangedListener 提供的 onTimeChanged() 方法获取。

控件 DatePicker 和 TimePicker 均包含了内部监听器接口，其定义如图 3.4.1 所示。

注意：实际使用这两个控件时，需要使用标准的 Java 日历类 java.util.Calendar 且月份值是 0～11。

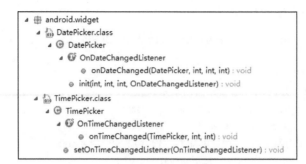

图 3.4.1　日期和时间选择器类的定义

3.4.2　自动完成文本控件 AutoCompleteTextView

自动完成文本控件 AutoCompleteTextView 用于实现文本的快速输入，其原理是事先将用于输入的文本存放在一个字符数组里，根据输入字符前方一致进行匹配（匹配字符个数由控件属性 completionThreshold 进行设置）。作为实用类的 AutoCompleteTextView 控件，其定义如图 3.4.2 所示。

图 3.4.2　自动完成文本控件的定义

3.4.3　标题栏 Toolbar 与 OptionMenu 菜单设计

菜单也是人机交互的重要方式。目前的 Android 手机软件一般不使用菜单键，而是推荐使用 Toolbar 代替默认的标题栏并创建 OptionMenu 菜单。此时，工具栏 Toolbar 右方区域会自动生成三个竖直点图形的菜单按钮 。

注意：Android 早期开发时，菜单设计针对手机上的菜单键。如今，原来的菜单键作为任务键了。

【例 3.4.1】　在标题栏 Toolbar 里创建 OptionMenu 菜单。

程序运行效果如图 3.4.3 所示。

图 3.4.3　Option Menu 菜单效果

为 Toolbar 创建 OptionMenu 菜单的主要步骤如下：

第 3 章　Android UI 与 Activity 组件

(1) 将菜单（项）保存在一个 .xml 文件里，并保存在 res/menu 文件夹里，示例代码如下：

```xml
<?xml version="1.0" encoding="utf-8"?>
<menu xmlns:android="http://schemas.android.com/apk/res/android">
    <item android:id="@+id/scan"
          android:icon="@android:drawable/ic_menu_myplaces"
          android:title="我的好友" />
    <item android:id="@+id/discoverable"
          android:icon="@android:drawable/ic_menu_view"
          android:title="设置在线" />
    <item android:id="@+id/back"
          android:icon="@android:drawable/ic_menu_close_clear_cancel"
          android:title="退出" />
</menu>
```

(2) 在主 Activity 的布局文件里定义 Toolbar 控件，其代码框架如下：

```xml
<!--新版 Android 支持的 Toolbar，并对标题栏布局，Toolbar 包含的布局及控件代码-->
<android.support.v7.widget.Toolbar
    android:id="@+id/toolbar"
    android:layout_width="match_parent"
    android:layout_height="wrap_content">
    <LinearLayout
        android:layout_width="match_parent"
        android:layout_height="match_parent"
        android:orientation="horizontal">
        <TextView
            android:id="@+id/title_left_text"
            style="?android:attr/windowTitleStyle"
            android:layout_width="0dp"
            android:layout_height="match_parent"
            android:layout_alignParentLeft="true"
            android:layout_weight="1"
            android:gravity="left"
            android:ellipsize="end"
            android:singleLine="true"
            android:text="蓝牙 Demo"/>
        <TextView
            android:id="@+id/title_right_text"
            android:layout_width="0dp"
            android:layout_height="match_parent"
            android:layout_alignParentRight="true"
            android:layout_weight="1"
            android:ellipsize="end"
            android:gravity="right"
            android:singleLine="true"
```

```
            android:text="连接至 OPPO R9s"/>
    </LinearLayout>
</android.support.v7.widget.Toolbar>
```

（3）在 Activity 组件程序的 onCreate()方法里，定义呈现 Toolbar 并进行选项单击事件处理的代码，其代码框架如下：

```
getSupportActionBar().hide();    //隐藏默认的标题栏
Toolbar toolbar = findViewById(R.id.toolbar);
toolbar.inflateMenu(R.menu.option_menu);    //呈现选项菜单
toolbar.setOnMenuItemClickListener(new MyMenuItemClickListener());
//选项菜单的单击事件处理
private class MyMenuItemClickListener implements Toolbar.OnMenuItemClickListener {
    @Override
    public boolean onMenuItemClick(MenuItem item) {
        switch (item.getItemId()) {
            case R.id.scan:
                Toast.makeText(MainActivity.this, "你点击了（我的好友）",
                    Toast.LENGTH_SHORT).show();
                return true;
            case R.id.discoverable:
                Toast.makeText(MainActivity.this, "你点击了（设置在线）",
                    Toast.LENGTH_SHORT).show();
                return true;
            case R.id.back:
                finish();
                System.exit(0);
                return true;
        }
        return false;
    }
}
```

3.4.4 Fragment 与 ListFragment

Fragment 系碎片之意。在 Android 中，Activity 组件界面可以由不同的 Fragment 组成。Fragment 有自己的生命周期和接收、处理用户的事件，这样就不必在一个 Activity 里面写一堆事件、控件的代码了。更为重要的是，可以动态地添加、替换、移除某个 Fragment。

使用 Fragment 的好处：

（1）因为一个 Fragment 可以被多个 Activity 嵌套，所以，能实现代码复用；

（2）Activity 用来管理 Fragment。Fragment 的生命周期寄托到 Activity 中，Fragment 可以被添加或释放。

使用 Fragment 时，需要在 Activity 组件的布局文件里使用标签<fragment>，其 name

第 3 章　Android UI 与 Activity 组件

属性为某个 Fragment 程序。类 ListFragment 作为展示列表数据程序的基类。

【例 3.4.2】 Fragment 布局及控件的使用。

一个使用 Fragment 布局及控件实现的模块运行效果如图 3.4.4 所示。

图 3.4.4　模块主要文件及运行效果

主 Activity 采用水平线性布局，分为左右两块。左边为一个 Fragment 控件，其 name 属性是用于显示书名称列表的 Fragment 程序；右边为一个 FrameLayout 布局，用于显示书详细信息。文件 activity_main.xml 的代码如下：

```xml
<?xml version="1.0" encoding="utf-8"?>
<LinearLayout
    xmlns:android="http://schemas.android.com/apk/res/android"
    android:layout_width="match_parent"
    android:layout_height="match_parent"
    android:layout_marginLeft="16dp"
    android:layout_marginRight="16dp"
    android:showDividers="middle"
    android:divider="?android:attr/dividerHorizontal"
    android:orientation="horizontal">
    <!-- 在水平排列的 LinearLayout 里，产生控件之间的分隔线，需要设置如下：
        showDividers="middle"：并指定在组件之间使用分隔条
        divider="?android:attr/dividerHorizontal"：设置线条
    -->
    <!-- 添加一个 Fragment，name 属性值关联一个带包名的 Fragment 程序 -->
    <fragment
        android:name="com.example.example3_6.BookListFragment"
        android:id="@+id/book_list"
        android:layout_width="0dp"
        android:layout_height="match_parent"
        android:layout_weight="1" />
    <!-- 添加一个 FrameLayout 容器，最终显示具体的 Fragment。-->
    <FrameLayout
        android:id="@+id/book_detail_container"
        android:layout_width="0dp"
```

```
        android:layout_height="match_parent"
        android:layout_weight="3" />
</LinearLayout>
```

布局 fragment_book_detail 采用垂直线性布局，包含两个 TextView 控件，分别显示书的标题和内容描述。

类 BookContent 定义了列表数据，并以模型类 Book 为内部类，其代码如下：

```
public class BookContent {
    public static class Book {    //定义内部类——模型类——系统的业务对象
        public Integer id;
        public String title;
        public String desc;
        public Book(Integer id, String title, String desc) {
            this.id = id;
            this.title = title;
            this.desc = desc;
        }
        @Override
        public String toString() {
            return title;
        }
    }
    // 定义外部类 BookContent 的 2 个公有属性
    // 1.List 集合用于显示书中内容
    public static List<Book> ITEMS = new ArrayList<Book>();
    // 2.Map 集合用于显示书中列表
    public static Map<Integer, Book> ITEM_MAP = new HashMap<Integer, Book>();
    static { // 使用静态初始化代码，调用成员方法
        addItem(new Book(1, "Android 应用开发案例教程", "一本深入介绍 android 开发架构的教
                                                                    材。"));
        addItem(new Book(2, ".NET 架构", "计算机专业选修课程 "));
        addItem(new Book(3, "轻量级 Java EE 企业应用实战", "全面介绍 Java EE 开发框架：
                                                    1.Struts 2.Spring 3.Hibernate"));
    }
    private static void addItem(Book book) {      //本方法供上面的静态代码块调用
        ITEMS.add(book);                          //添加到 List 集合
        ITEM_MAP.put(book.id, book);              //添加到 Map 集合
    }
}
```

类 BookListFragment 定义了内部接口 CallBacks，以便与其所属的 Activity 通信，其结构如图 3.4.5 所示。

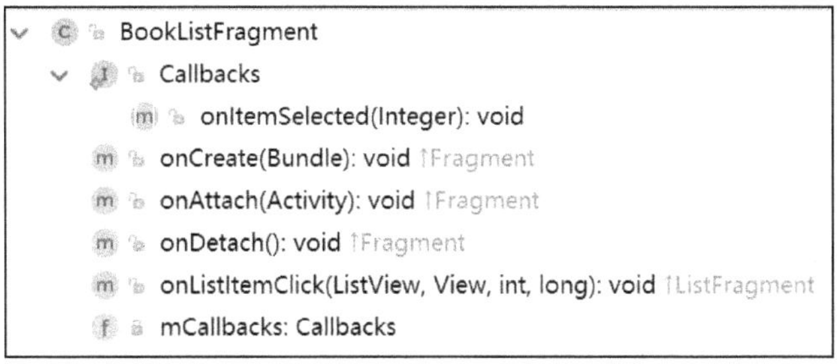

第 3 章 Android UI 与 Activity 组件

```
v C  BookListFragment
  v J  Callbacks
      m  onItemSelected(Integer): void
    m  onCreate(Bundle): void ↑Fragment
    m  onAttach(Activity): void ↑Fragment
    m  onDetach(): void ↑Fragment
    m  onListItemClick(ListView, View, int, long): void ↑ListFragment
    f  mCallbacks: Callbacks
```

图 3.4.5　类 BookListFragment 结构

类 BookListFragment 用于书的目录信息，作为 android.app.ListFragment 的子类，其代码如下：

```
/*
    使用 ListFragment 处理列表
    定义一个回调（内部）接口作为类成员，Fragment 所在 Activity 需要实现该接口
    Fragment 将通过该接口与它所在的 Activity 交互
*/
public class BookListFragment extends ListFragment {
    public interface Callbacks {                    //定义内部接口
        public void onItemSelected(Integer id);     //选择了某项时
    }
    private Callbacks mCallbacks;
    @Override
    public void onCreate(Bundle savedInstanceState) {
        super.onCreate(savedInstanceState);
        // 为该 ListFragment 设置 Adapter
        setListAdapter(new ArrayAdapter<BookContent.Book>(getActivity(),
                android.R.layout.simple_list_item_activated_1, android.R.id.text1,
                BookContent.ITEMS));    //使用公有属性 BookContent.ITEMS（不推荐）
    }
    //当该 Fragment 被附加、显示到 Activity 时，回调该方法
    @Override
    public void onAttach(Activity activity) {
        super.onAttach(activity);
        // 如果 Activity 没有实现 Callbacks 接口，抛出异常
        if (!(activity instanceof Callbacks)) {
            throw new IllegalStateException("BookListFragment 所在的 Activity 必须实现
                                                                    Callbacks 接口!");
        }
        // 把该 Activity 当成 Callbacks 对象
        mCallbacks = (Callbacks) activity;
    }
    //当该 Fragment 从它所属的 Activity 中被剥离时回调该方法
    @Override
    public void onDetach() {
```

```java
        super.onDetach();
        // 将 mCallbacks 赋为 null
        mCallbacks = null;
    }
    //当用户单击某列表项时激发该回调方法
    @Override
    public void onListItemClick(ListView listView, View view, int position, long id) {
        super.onListItemClick(listView, view, position, id);
        mCallbacks.onItemSelected(BookContent.ITEMS.get(position).id);
    }
}
```

类 BookDetailFragment 作为 Fragment 的子类,用于显示书的详细信息,其代码如下:

```java
public class BookDetailFragment extends Fragment {
    public static final String ITEM_ID = "item_id";
    BookContent.Book book;   //保存该 Fragment 显示的 Book 对象
    @Override
    public void onCreate(Bundle savedInstanceState) {
        super.onCreate(savedInstanceState);
        // 如果启动该 Fragment 时包含了 ITEM_ID 参数
        if (getArguments().containsKey(ITEM_ID)) {
            book = BookContent.ITEM_MAP.get(getArguments().getInt(ITEM_ID));
        }
    }
    @Override
    public View onCreateView(LayoutInflater inflater, ViewGroup container, Bundle savedInstanceState) {
        //加载书详情布局文件,包含标题和描述两个字段
        View rootView = inflater.inflate(R.layout.fragment_book_detail, container, false);
        if (book != null) {
            // 让 book_title 文本框显示 book 对象的 title 属性
            ((TextView) rootView.findViewById(R.id.book_title)).setText(book.title);
            // 让 book_desc 文本框显示 book 对象的 desc 属性
            ((TextView) rootView.findViewById(R.id.book_desc)).setText(book.desc);
        }
        return rootView;
    }
}
```

Activity 组件实现接口 BookListFragment.Callbacks,调用 FragmentManager 完成 Fragment 替换,其代码如下:

```java
/*
    MainActivity 实现了内部接口 BookListFragment.Callbacks
    使用 FragmentMananager 对象管理管理 Activity 碎片(Fragment)
*/
public class MainActivity extends Activity implements BookListFragment.Callbacks {
    @Override
```

```java
public void onCreate(Bundle savedInstanceState) {
    super.onCreate(savedInstanceState);
    setContentView(R.layout.activity_main);
}
@Override
public void onItemSelected(Integer id) {     //实现 Callbacks 接口方法
    //创建 Bundle,准备向 Fragment 传入参数
    Bundle bundle = new Bundle();
    bundle.putInt(BookDetailFragment.ITEM_ID, id);
    //创建 BookDetailFragment 对象
    BookDetailFragment fragment = new BookDetailFragment();
    //向 Fragment 传入参数
    fragment.setArguments(bundle);
    //使用 fragment 替换 book_detail_container 容器当前显示的 Fragment
    getFragmentManager().beginTransaction()
            .replace(R.id.book_detail_container, fragment)
            .addToBackStack(null)
            .commit();
}
```

【例 3.4.2】另法。

【知识要点】 与前面的方法相比,本法进行了两点优化:一是使用 Android 提供的 Application 类,实现 Activity 与 Fragment 之间的数据共享;二是不再使用回调接口。按照如下步骤改写后,代码更加清晰、易懂。

(1) 创建从类 BookContent 分离出来的实体类文件 model.Book.java。

(2) 定义 Application 的子类 BookContent,其代码如下:

```java
public class BookContent extends Application {
    private List<Book> books ;                        //类成员属性
    public List<Book> getBooks() { return books;     //供 BookListFragment 调用}
    public void setBooks(List<Book> books) { this.books = books; }
    @Override
    public void onCreate() {
        super.onCreate();
        List<Book> books=new ArrayList<>();
        books.add(new Book(1, "Android 应用开发案例教程",
                            "一本深入介绍 android 开发架构的教材。"));
        books.add(new Book(2, ".NET 架构","计算机专业专业选修课程 "));
        books.add(new Book(3, "轻量级 Java EE 企业应用实战", "全面介绍 Java EE 开发框架:
                            1.Struts 2.Spring 3.Hibernate"));
        setBooks(books);                             //完成对象创建
    }
    public Book findBook(int bookId) {                //在类 BookDetailFragment 里使用
        for (Book book : books)
            if (book.getId()==bookId)   return   book;
```

```
            return null;
    }
}
```

(3) 在清单文件的<application>标签里,使用如下代码对类 BookContent 进行注册:

`android:name=".model.BookContent"`

(4) 在类 BookListFragment 里定义 BookContent 类型的属性 bookContent (对象),并在其 onCreate()方法里,通过如下方式实例化:

`bookContent=(BookContent) getActivity().getApplication();`
`//bookContent=(BookContent) getActivity().getApplicationContext();    //也 OK`

在相关方法里使用 bookContent.getBooks()方式获取 books 对象(书籍列表)。

(5) 在类 BookDetailFragment 的 onCreate()方法里,也是使用 BookContent 类型的对象 bookContent,然后调用 findBook()方法获取书籍的详细信息。

(6) 在 BookListFragment 重写的方法 onListItemClick()里,使用 getFragmentManager() 方法获取当前 ListFragment 对象的父容器的 FragmentManager 对象。这种方式,就不需要在类 BookListFragment 里写回调接口、在类 MainActivity 里写实现代码。

3.4.5 底部导航 BottomNavigationView

android.support.design.widget.BottomNavigationView 是新版 Android 增加的一个控件,用于设计手机的底部导航效果。在 Android Studio 3.1.2 里,新建项目(模块)时,选择布局为 Button Navigation Activity,默认创建的底部导航效果如图 3.4.6 所示。

图 3.4.6 底部导航效果示例

控件 BottomNavigationView 位于主 Activity 的布局文件里,并与一个展示 Fragment 内容的 FrameLayout 布局联合使用,其代码如下:

```xml
<?xml version="1.0" encoding="utf-8"?>
<android.support.constraint.ConstraintLayout xmlns:android="http://schemas.android.com/apk/res/android"
    xmlns:app="http://schemas.android.com/apk/res-auto"
    xmlns:tools="http://schemas.android.com/tools"
    android:id="@+id/container"
    android:layout_width="match_parent"
    android:layout_height="match_parent"
    tools:context=".MainActivity">
    <FrameLayout
        android:id="@+id/fragment_content"
```

```xml
            android:layout_width="match_parent"
            android:layout_height="0dp"
            android:layout_weight="1">
    </FrameLayout>
    <TextView
        android:id="@+id/message"
        android:layout_width="wrap_content"
        android:layout_height="wrap_content"
        android:layout_marginLeft="92dp"
        android:layout_marginTop="@dimen/activity_vertical_margin"
        android:text="@string/title_home"
        app:layout_constraintLeft_toLeftOf="parent"
        app:layout_constraintTop_toTopOf="parent" />
    <android.support.design.widget.BottomNavigationView
        android:id="@+id/navigation"
        android:layout_width="match_parent"
        android:layout_height="79dp"
        android:background="?android:attr/windowBackground"
        app:layout_constraintBottom_toBottomOf="parent"
        app:layout_constraintHorizontal_bias="0.0"
        app:menu="@menu/navigation" />
</android.support.constraint.ConstraintLayout>
```

控件 BottomNavigationView 的一个重要属性是 app:menu，用以引用存放在 res/menu 里的菜单文件，其代码如下：

```xml
<?xml version="1.0" encoding="utf-8"?>
<menu xmlns:android="http://schemas.android.com/apk/res/android">
    <!--每个菜单项包含 id、图标和菜单文本三项-->
    <item
        android:id="@+id/navigation_home"
        android:icon="@drawable/ic_home_black_24dp"
        android:title="@string/title_home" />
    <item
        android:id="@+id/navigation_dashboard"
        android:icon="@drawable/find"
        android:title="@string/title_dashboard" />
    <item
        android:id="@+id/navigation_notifications"
        android:icon="@drawable/ic_notifications_black_24dp"
        android:title="@string/title_notifications" />
</menu>
```

在 MainActivity 组件里，监听对菜单项的选择，并切换到相应的 Fragment 视图，其代码如下：

```java
public class MainActivity extends AppCompatActivity {
    private TextView mTextMessage;
```

```java
        private FragmentManager fragmentManager;
        private OneFragment oneFragment;
        private BlankFragment blankFragment;
        private FragmentTransaction transaction;
        public static String static_nikenameAtMainActivity;
        public static String static_phoneNumberAtMainActivity;
        private BottomNavigationView.OnNavigationItemSelectedListener mOnNavigationItemSelectedListener
                        = new BottomNavigationView.OnNavigationItemSelectedListener() {
            @Override
            public boolean onNavigationItemSelected(@NonNull MenuItem item) {
                fragmentManager = getFragmentManager();
                transaction = fragmentManager.beginTransaction();
                int id = item.getItemId();
                switch (id) {
                    case R.id.navigation_home: {
                        setTitle("home ");
                        transaction.replace(R.id.fragment_content, oneFragment);
                        //要求引 Fragment 与引 FragmentManager 的包一致!
                        transaction.commit();
                        return true;
                    }
                    case R.id.navigation_dashboard: {
                        setTitle("dashboard ");
                        transaction.replace(R.id.fragment_content,blankFragment);
                        transaction.commit();
                        return true;
                    }
                    case R.id.navigation_ notifications: {
                        setTitle("notifications ");
                        transaction.replace(R.id.fragment_content,blankFragment);
                        transaction.commit();
                        return true;
                    }
                    default: return    false;
                }
            }
        };
        @Override
        protected void onCreate(Bundle savedInstanceState) {
            super.onCreate(savedInstanceState);
            setContentView(R.layout.activity_main);
            mTextMessage = (TextView) findViewById(R.id.message);
            oneFragment = new OneFragment();//1
            blankFragment=new BlankFragment();
            BottomNavigationView navigation = (BottomNavigationView) findViewById(R.id.navigation);
            navigation.setOnNavigationItemSelectedListener (mOnNavigationItemSelectedListener);
        }
    }
```

习 题 3

一、判断题

1. 在 Android 应用中，界面布局有图形和代码两种设计方式。
2. 所有界面元素都是 View 类的子类。
3. 在一个布局文件里，所有元素的 ID 都可以重复。
4. Context 是 Activity 的超类。
5. AlertDialog 位于软件包 android.widget 内。
6. 布局标签不能嵌套。
7. 线性布局就是控件对象的水平排列。
8. 配置项目的主 Activity 时，使用了意图过滤器标签<intent-filter>。
9. 弹出 Toast、启动 Activity 或 Service、发送广播和操作数据库等，都需要使用 ConText 对象。

二、选择题

1. 下列 Android 事件中，其回调方法没有返回值的是_____。
   A．OnClick　　　B．OnKeyDown　　　C．OnTouch　　　D．OnFocusChange
2. 在 Android Studio 3.1.2 中，创建 Android 项目时，默认使用的布局类型是_____。
   A．相对布局　　　B．帧布局　　　C．表格布局　　　D．约束布局
3. 在 Android 布局文件中，新建一个资源共享 ID 的正确方法是_____。
   A．android:id="id/name"　　　　　B．android:id="@id/name"
   C．android:id="@+id/name"　　　　D．android:id="@id+/name"
4. 在使用 RadioButton 时，要想实现互斥的选择需要_____。
   A．ButtonGroup　B．RadioButtons　C．CheckBox　D．RadioGroup
5. 下列选项中，不能使用 UI 图形化设计方式的是_____。
   A．TextView　　　B．EditText　　　C．Toast　　　D．ImageView

三、填空题

1. Activity 组件最常用的生命周期方法是_____。
2. 在 Activity 组件程序里，设置 ImageView 控件来源所使用的方法是_____。
3. 在布局文件中，设置控件大小通常使用 dp 作为单位，而设置文字大小的单位是_____。
4. 对于处于运行态的 Activity，当用户按返回键退出时，最后调用的生命周期方法是_____。
5. ListView 控件的适配器必须是_____接口类型。

实 验 3

一、实验目的

(1) 掌握常用的布局方法。
(2) 掌握 Android 视图类 View 的使用,特别是单击事件监听器和动态布局的使用。
(3) 掌握 Activity 组件的生命周期。
(4) 掌握 Android 常用控件的使用。
(5) 掌握在标题栏 Toolbar 中创建 OptionMenu 的方法。
(6) 掌握 Fragment 布局及控件的使用。
(7) 掌握底部导航的使用。

二、实验内容及步骤

【预备】在 Android Studio 中,新建名为 Example3 的项目后,访问本课程配套的网站 http://www.wustwzx.com/as/sy/sy03.html,复制相关代码,完成如下几个模块的设计。

1. 掌握各种布局的特点、Android 常用控件的使用(参见例 3.1.1)

(1) 在项目里,新建名为 example3_1 的模块。
(2) 在默认的约束布局里,添加垂直线性布局并内嵌水平线性布局,然后依次添加文本框、下拉列表、单选按钮组和复选框等控件。
(3) 部署模块并做运行测试。
(4) 在文件夹 res/layout 里新建一个布局文件,指定根标签为 FrameLayout(帧布局)。在 Design 模式下,向该布局依次添加两个控件,查验后添加的控件对象会在前一控件对象上进行覆盖而形成遮挡。

2. 掌握 Activity 组件的生命周期

(1) 在项目里,新建名为 example3_2 的模块,在 MainActivity 程序 onCreate()方法里,使用 Log.i()语句,打印一条 Log 信息。
(2) 在 onCreate()方法体外的空白处,按【Ctrl+O】组合键,选择 onRestart(),在该方法里也使用 Log.i()语句打印一条 Log 信息,其 Tag 名与 onCreate()方法里 Log.i()语句指定的名称相同。依次重写组件的其他生命周期方法 onStart()、onResume()、onPause()、onStop()和 onDestory()。
(3) 部署本模块并运行,打开 Logcat 控制台,按照程序里使用的 Tag 名过滤,观察所执行的生命周期方法。
(4) 按返回键,观察所执行的生命周期方法。
(5) 再次运行,将手机切换成横屏,观察所执行的生命周期方法。查验在执行 onDestory()后,再重新执行 onCreate()等生命周期方法。

第 3 章　Android UI 与 Activity 组件

3. 在对话框里动态加载布局视图（参见例 3.3.1）

（1）在项目里，新建名为 example3_3 的模块。在文件夹 res/layout 里，创建一个用于登录的布局文件 login_view.xml，它主要包含两个 EditText 控件。

（2）在主布局文件里，添加一个名为 btn 的 Button 控件。

（3）在 Activity 组件里，创建一个单击监听器对象 listener 作为类成员，在其重写的方法 onClick()里，先使用 LayoutInflater 对象加载布局而得到一个 View 对象，然后再创建一个 AlertDialog 类型的对话框，设置视图为刚才得到的 View 对象。最后，完成对话框里按钮的单击事件的业务逻辑。

（4）在 Activity 组件的 onCreate()方法里，对按钮 btn 应用单击事件监听器 listener。

（5）部署本模块并做运行测试。

4．ListView 控件及数据适配器（含列表项监听器）的使用（参见例 3.3.2）

（1）在项目里，新建名为 example3_4 的模块，在主布局文件里添加 3 个 Button 按钮和 1 个 ListView 控件。

（2）修改类 MainActivity，让其实现接口 View.OnClickListener，重写接口方法；将上述控件对象作为类成员；在 onCreate()方法内，分别注册 3 个按钮的单击事件监听器。

（3）在文件夹 res/layout 内，创建供适配器 ArrayAdapter 使用的布局文件；在第 1 个按钮的监听器代码内，创建适配器 ArrayAdapter 对象，并应用到 ListView 控件；最后设置 ListView 控件的列表项单击监听器。

（4）在文件夹 res/layout 内，创建供适配器 SimpleAdapter 使用的布局文件；在第 2 个按钮的监听器代码内，创建适配器 SimpleAdapter 对象，并应用到 ListView 控件；最后设置 ListView 控件的列表项单击监听器。

（5）复制 4 个人物图片文件至文件夹 res/drawable；在文件夹 res/layout 内，创建供适配器 BaseAdapter 使用的布局文件；在第 3 个按钮的监听器代码内，创建适配器 BaseAdapter 对象，并应用到 ListView 控件；最后设置 ListView 控件的列表项单击监听器。

（6）部署本模块并做运行测试。

5．在标题栏 Toolbar 里创建 OptionMenu 菜单（参见例 3.4.1）

（1）在项目里，新建名为 example3_5 的模块，编写菜单（项）文件并保存在文件夹 res/menu 里。

（2）在 Activity 组件的布局文件里，添加控件 android.support.v7.widget.Toolbar。

（3）在 Activity 组件程序的 onCreate()方法里，定义呈现 Toolbar 并进行选项单击事件处理的代码。

（4）部署本模块并做运行测试。

6．Fragment 布局及控件的使用（参见例 3.4.2）

（1）在项目里，新建名为 example3_6 的模块。在布局文件里，水平放置<fragment>

和< frameLayout >两个标签，并在<fragment>标签里指定相应的 ListFragment 程序。

（2）在程序文件夹里，创建列表数据及其对应的实体类；编写 ListFragment 程序代码，其内定义了与 Activity 组件通信的内部接口。

（3）在文件夹 res/layout 里，创建供显示详细信息的布局文件；编写供 Activity 组件调用、显示详细内容的 Fragment 程序代码。

（4）让 MainActivity 实现 ListFragment 的内部接口，重写接口方法，实现 Fragment 的动态更新。

（5）部署本模块并做运行测试。

三、实验小结及思考

（由学生填写，重点写上机中遇到的问题。）

第 4 章 手机基本功能程序设计

如今,手机不再仅仅作为打电话和发短信的工具。事实上,音频和视频的播放与录制、二维码扫描等功能,也深受手机用户的喜爱。本章主要介绍 Android 手机常用功能的程序设计,其学习要点如下:
- 掌握带数据传递的 Activity 组件调用与返回;
- 掌握 Android 6.0 动态权限模型;
- 掌握 Android 共享存储与文件存储的方法;
- 掌握打电话程序的设计方法;
- 掌握短信发送程序的设计方法;
- 掌握手机音频播放程序、录音程序和视频播放程序的设计方法。

4.1 预 备 知 识

4.1.1 Activity 组件的调用与返回

1. 组件的显式调用与隐式调用

Activity 组件调用可分为显式调用和隐式调用。显式调用是通过组件名来完成的调用,示例代码如下:

```
Intent intent=new Intent();
intent.setClass(MainActivity.this,OtherActivity.class);   //组件显式调用
startActivity (intent);
```

隐式调用是通过意图过滤器(IntentFilter)来实现的,它没有明确指出目标组件的名称。Android 系统会根据隐式意图中设置的动作(Action)、类别(Category)、数据(URI 和数据类型)找到最合适的组件来处理这个意图,一般用于不同应用程序之间的调用。

IntentFilter 的过滤信息有 Action、Category 和 Data,只有同时匹配过滤列表中的信息,才可以匹配成功。一个 Activity 中可以有多个 IntentFilter,一个 Intent 只要有一组完全匹配,就可以成功启动对应的 Activity。

Action 匹配规则:Action 是一个区分大小写的字符串,要求必须存在且和过滤规则中的一个 Action 相同。

Category 匹配规则：添加<category android:name="android.intent.category.DEFAULT" />，这个 category 是使用隐式调用所必需的，否则，无法开启被调用的 Activity 组件。

Data 匹配规则：如果在 AndroidManifest.xml 里面指定了<data>这行，那么，若需要匹配到它，在代码里必须设置 Intent 的 Data 信息。

注意：

（1）通常情况下，在清单文件里使用标签<intent-filter>来配置意图过滤器。

（2）Android 系统必须在启动一个组件之前知道其所具有的功能。因此，活动、服务和广播接收者都可以拥有一个或多个意图过滤器,并以隐式调用方式来启动目标组件。

2．有数据传递和值返回的 Activity 调用

Activity 组件调用时可能存在数据的传递。此时，需要使用 Activity 具有的方法 startActivityForResult（Intent intent, int requestCode）去打开一个新的 Activity，新的 Activity 关闭后会向先前的 Activity 传回数据。为了得到传回的数据，必须在主调 Activity 组件里重写方法 onActivityResult（int requestCode, int resultCode, Intent data）。其中，使用 Intent 对象携带数据，其数据类型可以是 String 和 Bundle 等。

【例 4.1.1】 有参数传递和返回值的 Activity 调用。

在 MainActivity 组件程序对应的布局文件里，定义了两个 Button 按钮。单击按钮后，启动对应的 Activity 组件程序。调用其他主调组件的代码如下：

```java
/*
    通过 Intent 对象实现 Activity 组件的调用
    被调用的 Activity 组件：显式调用和隐式调用
    是否使用请求码：startActivity(intent)和 startActivityForResult(intent,requestCode)
    组件返回时使用结果码：setResult(resultCode,intent1)
    使用 Intent 携带数据：捆绑或未捆绑
*/
public class MainActivity extends AppCompatActivity {
    @Override
    protected void onCreate(Bundle savedInstanceState) {
        super.onCreate(savedInstanceState);
        setContentView(R.layout.activity_main);
        Button button1=findViewById(R.id.button);
        Button1.setOnClickListener(new View.OnClickListener() {
            @Override
            public void onClick(View v) {
                Intent intent=new Intent();
                //在清单文件中已配置了动作名称
                intent.setAction("android.intent.action.ysdy" );   //隐式调用
                startActivityForResult(intent,2);    //第 2 个参数为请求码
            }
        });
        Button button2=findViewById(R.id.button2);
```

```java
        button2.setOnClickListener(new View.OnClickListener() {
            @Override
            public void onClick(View v) {
                Intent intent=new Intent();
                //简单用法
                intent.setClass(MainActivity.this,ThirdActivity.class);    //组件显式调用
                /*//其他写法
                ComponentName componentName = new ComponentName(
                                                MainActivity.this, ThirdActivity.class);
                intent.setComponent(componentName);*/
                Bundle bundle=new Bundle();
                //键值对数据存入 Bundle 对象
                bundle.putString("name","张三");bundle.putInt("age",20);
                //通过意图对象携带数据
                intent.putExtra("data",bundle);
                startActivityForResult(intent,3);       //第 2 个参数为请求码
            }
        });
    }
    @Override
    protected void onActivityResult(int requestCode, int resultCode, Intent data) {
        super.onActivityResult(requestCode, resultCode, data);
        //判断从哪个 Activity 返回,使用请求码
        Toast.makeText(this, "从第"+requestCode+"个 Activity 返回。", Toast.LENGTH_LONG).
                                                                                    show();
        //判断是哪个 Activity 返回的数据,使用结果码
        if(resultCode==3){
            String string=data.getStringExtra("hello");    //获取未捆绑的返回结果数据
            Toast.makeText(this, string, Toast.LENGTH_LONG).show();
        }
    }
}
```

其中,在清单文件里,对被调组件的配置代码如下:

```xml
<activity
    android:name=".SecondActivity"
    android:label="隐式调用 SecondActivity">
    <intent-filter>
        <action android:name="android.intent.action.ysdy" />
        <category android:name="android.intent.category.DEFAULT" />
    </intent-filter>
</activity>
<activity android:name=".ThirdActivity"></activity>
```

第 2 个 Activity 组件的代码如下:

```java
public class SecondActivity extends AppCompatActivity {
```

```java
    @Override
    protected void onCreate(Bundle savedInstanceState) {
        super.onCreate(savedInstanceState);
        setContentView(R.layout.activity_second);
        Button button = findViewById(R.id.button);
        button.setOnClickListener(new View.OnClickListener() {
            @Override
            public void onClick(View v) {
                finish();
                Log.i("wutest", "第 2 个 Activity 准备销毁...: ");
            }
        });
    }
    @Override
    protected void onDestroy() {
        super.onDestroy();
        Log.i("wutest", "第 2 个 Activity 销毁了。");
        //当前 Activity 销毁后，返回父 Activity。
        //如果在主 Activity 中使用 finish()方法，将退出应用，其作用如同按返回键。
    }
}
```

第 3 个 Activity 组件接收传递的捆绑数据，返回前设置非捆绑数据和结果码，其代码如下：

```java
/*
    接收从 MainActivity 传递的 Bundle 数据
    销毁前，向主 Activity 传递数据
*/
public class ThirdActivity extends AppCompatActivity {
    @Override
    protected void onCreate(Bundle savedInstanceState) {
        super.onCreate(savedInstanceState);
        setContentView(R.layout.activity_third);
        String receiver="接收的数据如下：\n";
        Intent intent=getIntent();
        Bundle data = intent.getBundleExtra("data");      //获取捆绑数据
        receiver+="name："+data.getString("name")+"\n";
        receiver+="age："+data.getInt("age");
        Toast.makeText(this, receiver, Toast.LENGTH_LONG).show();
        intent=new Intent();
        intent.putExtra("hello","How are you?");   //携带未捆绑的数据
        setResult(3,intent);    //第 1 个参数是设置结果码
    }
}
```

注意：供 MainActivity 组件调用的其他 Activity 组件，均需要在清单文件里注册。

事实上，使用菜单 File→New→Activity 创建时，该 Activity 组件会自动在清单文件里使用标签<activity>注册。

4.1.2 Android 权限、权限组与运行时权限动态检测

权限是一种安全机制。Android 权限主要用于限制应用程序内部某些具有限制特性的功能使用。例如，当 Android 应用有联网需求时，就应当在项目清单文件里添加如下代码：

```
<uses-permission android:name="android.permission. INTERNET"/>
```

新的 Android 安全架构将权限划分为普通权限和危险权限。普通权限是指不会对用户隐私或设备操作造成很大风险的权限，如手机振动、访问网络等，由于它们在运行时只需要在清单文件里申请，因此系统在运行时会自动授予这些权限。

危险权限是指可能泄露用户隐私或影响设备正常操作的权限，如访问外部存储的文件、访问通讯录等，Android 系统会要求用户明确授予这些权限，发出权限请求的方式取决于系统版本。

如果设备运行的是 Android 6.0（API 23）或更高版本，并且应用的 targetSdkVersion 是 23 或更高版本，则应用将在运行时向用户请求权限(Runtime Permissions)。用户可随时撤销权限，因此应用每次运行时都应该检查自身是否具备所需的权限。

如果设备运行的是 Android 5.1（API 22）或更低版本，并且应用的 targetSdkVersion 是 22 或更低版本，则系统在用户安装应用时就要求用户授予权限。用户一旦安装应用，撤销权限的唯一方式就是卸载应用。

Android 6.0 为了保护用户隐私等，对危险权限在应用运行时进行检测，用户可以选择允许、提醒和拒绝。

类 android.Manifest 定义了权限组及权限两个静态内部类，如图 4.1.1 所示。

```
public final class Manifest {
    public static final class permission_group {
        public static final String CALENDAR = "android.permission-group.CALENDAR";
        public static final String CAMERA = "android.permission-group.CAMERA";
        public static final String CONTACTS = "android.permission-group.CONTACTS";
        public static final String LOCATION = "android.permission-group.LOCATION";
        public static final String MICROPHONE = "android.permission-group.MICROPHONE";
        public static final String PHONE = "android.permission-group.PHONE";
        public static final String SENSORS = "android.permission-group.SENSORS";
        public static final String SMS = "android.permission-group.SMS";
        public static final String STORAGE = "android.permission-group.STORAGE";
    }
    public static final class permission {
        public static final String ACCESS_CHECKIN_PROPERTIES = "android.permission.ACCESS_CHECKIN_PROPERTIES";
        public static final String ACCESS_COARSE_LOCATION = "android.permission.ACCESS_COARSE_LOCATION";
        public static final String ACCESS_FINE_LOCATION = "android.permission.ACCESS_FINE_LOCATION";
        ......
    }
}
```

图 4.1.1 类 Manifest 的两个内部类 permission_group 和 permission

在 Android Studio 的 Terminal 控制台，分组形式显示 Android 所有危险权限的命令

格式如下：

adb shell pm list permissions -d -g

其中，-d 表示危险权限；-g 表示对权限进行分组。

显示 Android 所有危险权限组的命令如下：

adb shell pm list permission-groups -d

Android 权限组包含的常用危险权限如表 4.1.1 所示。

表 4.1.1 Android 危险权限（组）

权 限 组 名	权 限
android.permission-group.SMS （短信）	android.permission.READ_SMS
	android.permission.RECEIVE_SMS
	android.permission.SEND_SMS
android.permission-group.LOCATION （定位）	android.permission.ACCESS_FINE_LOCATION
	android.permission.ACCESS_COARSE_LOCATION
android.permission-group.PHONE （电话）	android.permission.CALL_PHONE
	android.permission. READ_PHONE_STATE
	android.permission. READ_CALL_LOG
android.permission-group.STORAGE （外部存储）	android.permission.READ_EXTERNAL_STORAGE
	android.permission.WRITE_EXTERNAL_STORAGE
android.permission-group.CONTACTS （手机联系人）	android.permission.READ_CONTACTS
	android.permission.WRITE_CONTACTS
	android.permission.GET_ACCOUNTS
android.permission-group.CAMERA （相机）	android.permission.CAMERA
android.permission-group.MICROPHONE （麦克风）	android.permission.RECORD_AUDIO
android.permission-group.CALENDAR （日历）	android.permission.READ_CALENDAR
	android.permission.WRITE_CALENDAR
android.permission-group.SENSORS （传感器）	android.permission.BODY_SENSORS

危险权限分组的好处：在动态申请权限时，一旦授予了权限组中的某个权限，则同一组的其他危险权限都将被授权。

1．单个危险权限应用

一个应用，如果只涉及一个危险权限组，在 Activity 组件里动态申请的示例代码框架如下：

```
@Override
```

```java
protected void onCreate(Bundle savedInstanceState) {
    super.onCreate(savedInstanceState);
    setContentView(R.layout.activity_main);
    //电话权限
    if (ActivityCompat.checkSelfPermission(MainActivity.this, Manifest.permission.CALL_PHONE) !=
                            PackageManager.PERMISSION_GRANTED) {
        //没有权限时请求该权限：出现"是否型"对话框由用户选择是否授权，使用了请求码
        ActivityCompat.requestPermissions(MainActivity.this, new String[]{
                            Manifest.permission.CALL_PHONE}, 1);
    }else{
        fun();   //执行有权限时的业务逻辑，fun()需要用户定义
    }
}
void fun(){
    //有电话权限时的业务逻辑
}
@Override   //Android 6.0 动态权限处理的接口回调方法
public void onRequestPermissionsResult(int requestCode, @NonNull String[] permissions,
                            @NonNull int[] grantResults) {
    super.onRequestPermissionsResult(requestCode, permissions, grantResults);
    switch (requestCode) {
        case 1:
            if (grantResults[0] == PackageManager.PERMISSION_GRANTED) {   //所需权限
                fun();
            } else {
                Toast.makeText(this, "没有打电话权限", Toast.LENGTH_LONG).show();
                finish();    //销毁当前 Activity 组件
            }
    }
}
```

2. 多个危险权限组应用

一个应用，如果只涉及多个危险权限组，在 Activity 组件里，应先筛选未授权的危险权限到一个数组里，然后集中动态申请；在权限回调方法中，如果检测到有权限未授予，则会给出消息提示后销毁当前的 Activity 组件。一个包含有 3 个权限组的示例代码框架如下：

```java
//定义危险权限数组
private static String[] permissions = {"android.permission.READ_PHONE_STATE",
                            "android.permission.SEND_SMS",
                            "android.permission.WRITE_EXTERNAL_STORAGE"};
@Override
protected void onCreate(Bundle savedInstanceState) {
    super.onCreate(savedInstanceState);
    setContentView(R.layout.activity_main);
    //筛选未授权的权限
```

```java
        List<String> mPermissionList = new ArrayList<>();
        mPermissionList.clear();
        for (int i = 0; i < permissions.length; i++) {
            if (ContextCompat.checkSelfPermission(this, permissions[i]) !=
                                        PackageManager.PERMISSION_GRANTED) {
                mPermissionList.add(permissions[i]);
            }
        }
        //Build.VERSION.SDK_INT 表示 Android 设备的 API 版本
        //符号常量 Build.VERSION_CODES.M=23,对应于 Android 6.0
        if (Build.VERSION.SDK_INT < Build.VERSION_CODES.M) {   //Android 6.0 以下版本时
            fun();    //做该做的
        } else {
            if (mPermissionList.isEmpty()) {
                fun();   //应用已经获得权限时直接执行
            } else{
                //集中请求未授权的权限;将 List 转为数组
                String[] needPermissions = mPermissionList.toArray(new String[mPermissionList.size()]);
                //请求权限后执行回调方法
                ActivityCompat.requestPermissions(this, needPermissions, 1);
            }
        }
    }
    public void fun(){
        //有全部权限时的业务逻辑
    }
    @Override   //Android 6.0 动态权限处理的回调方法
    public void onRequestPermissionsResult(int requestCode, @NonNull String[] permissions,
                                        @NonNull int[] grantResults) {
        super.onRequestPermissionsResult(requestCode, permissions, grantResults);
        switch (requestCode) {
            case 1:
                for (int i = 0; i < grantResults.length; i++) {
                    if (grantResults[i] != PackageManager.PERMISSION_GRANTED) {
                        Toast.makeText(this, "权限不足,无法实现全部功能! ",
                                        Toast.LENGTH_SHORT).show();
                        finish();
                    }
                }
                fun();    //全部授权时执行
        }
    }
```

注意:

(1) 在 Android 组件程序里,如果没有重写 onRequestPermissionsResult()方法,在首次运行时,尽管赋予了所有权限,但不会立即生效,而以后运行时会因有权限而正常。

（2）Android 8.0 带来了全新的底层优化，对第三方 App 的权限要求更高，后台的活动性具有智能优化，系统也进行了一定程度的精简，力求达到更快速、更流畅、更安全和更省电的目的。

4.1.3 SharedPreferences 存储与文件存储

Android 文件存储分为内部存储（Internal Storage）与外部存储（External Storage）。

内部存储是将应用程序的数据以文件的方式保存至设备内存中，该文件为其创建的应用程序私有，其他应用程序无权进行操作，当该应用程序被卸载时，其内部存储文件也随之被删除。在重新安装时，先前创建、存储的文件不会被删除。

外部存储是指对 SD 卡里文件的读/写操作，除了在清单文件中要注册权限外，还需要在 Activity 组件程序里动态申请。

1. 使用 SharedPreferences 接口实现共享存储

为了方便用户操作，可以将用户名及密码保存至某个文件里，供下一次登录时使用，节省用户输入时间，这就需要使用 Android 提供的 SharedPreferences 接口。用户信息保存在项目包名下的 shared_prefs 文件夹的.xml 文件里，该文件保存在 data/data/包名/shared_prefs 里，以键值对的形式存放 Activity 文本框等的历史输入值。

SharedPreferences 接口位于软件包 android.content 里，每个 Activity 都有一个 SharedPreferences 接口类型的对象，该对象可以通过使用 Activity 超类 Context 提供的方法 getPreferences()得到。

SharedPreferences 接口提供了获取用户信息的方法 getString()及编辑用户信息的方法 edit()。edit()方法的返回值为内部接口 SharedPreferences.Editor 类型。内部接口 Editor 提供了以键值对形式保存用户信息的方法 putString()和提交用户信息方法 commit()等，如图 4.1.2 所示。

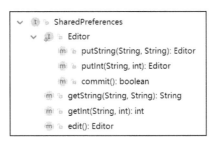

图 4.1.2 SharedPreferences 接口的内部接口

Android 应用程序默认安装至手机内部存储区，应用（模块）example3_2 会生成隶属于该应用的.xml 文件，存放在系统文件夹 data/data/pn/shared_prefs 里，通过 Android Studio 的 Device File Explorer 可以查看，如图 4.1.3 所示。

图 4.1.3　隶属于应用（模块）的私有文件

注意：SharedPreferences 本质上属于 Android 内部存储。

【例 4.1.2】　使用 SharedPreferences 接口实现的用户登录程序设计。

输入登录信息后，在单击"登录"按钮之前，可以勾选"保存登录信息"复选框，如图 4.1.4 所示。

图 4.1.4　用户的登录程序设计

Activity 组件程序的代码如下：

```
/*
    使用 SharedPreferences 实现偏好设置，可保存登录信息供下次登录用
*/
public class MainActivity extends AppCompatActivity {
    EditText et1, et2;   //两个文本框
    SharedPreferences sp;   //接口类型
    @Override
    protected void onCreate(Bundle savedInstanceState) {
        super.onCreate(savedInstanceState);
        setContentView(R.layout.activity_main);
        et1 = findViewById(R.id.edit1);
        et2 = findViewById(R.id.edit2);
        sp = getPreferences(Activity.MODE_PRIVATE);   //创建 SharedPreferences 对象
        String username = sp.getString("username", "");   //第 2 个参数表示按键名取不到值时,设
                                                                置为空串
        String password = sp.getString("password", "");
        et1.setText(username);
```

```java
            et2.setText(password);
        Button btn1=findViewById(R.id.btn1);
        btn1.setOnClickListener(new View.OnClickListener(){
            @Override
            public void onClick(View v) {
                String s1 = et1.getText().toString();
                String s2 = et2.getText().toString();
                //显示输入的登录信息
                Toast.makeText(getApplicationContext(), "输入的用户名为" + s1 + ",输入的密码
                                                                                 为" + s2,
Toast.LENGTH_LONG).show();
            }
        });
        Button btn2=findViewById(R.id.btn2);
        btn2.setOnClickListener(new View.OnClickListener(){
            @Override
            public void onClick(View v) {
                et1.setText("");
                et2.setText("");
                Toast.makeText(getApplicationContext(), "取消登录", Toast.LENGTH_SHORT).
                                                                                 show();
            }
        });
        CheckBox cb = findViewById(R.id.cb);      //找到复选框
        cb.setOnCheckedChangeListener(new CompoundButton.OnCheckedChangeListener()
                                                                       {    //选中监听
            @Override
            public void onCheckedChanged(CompoundButton arg0, boolean arg1) {
                if (arg1) { //勾选时
                    Toast.makeText(MainActivity.this, "已保存登录信息", Toast.LENGTH_
                                                                              LONG).show();
                    sp.edit().putString("username", et1.getText().toString())  //保存
                            .putString("password", et2.getText().toString())
                            .commit();
                } else {    //未勾选时
                    //第一个参数要求为 Context 类型,外部类对象
                    Toast.makeText(MainActivity.this, "不保存登录信息", Toast.LENGTH_
                                                                              LONG).show();
                    sp.edit().putString("username", null)     //不保存
                            .putString("password", null)
                            .commit();
                }
            }
        });
    }
}
```

2. Android 内部存储

抽象类 android.content.Context 定义了文件的操作模式常量；而作为类 Context 的包装类 android.content.ContextWrapper，它提供了用于文件读/写的方法 openFileOutput()和 openFileInput()，其类型对应于标准 I/O 文件读/写类型，如图 4.1.5 所示。

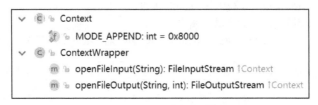

图 4.1.5　抽象类 Context 及其包装类 ContextWrapper

注意：

（1）在 Android 应用开发中，还需要配合使用 Java 中的标准 I/O 文件读/写方式来实现文件读/写。

（2）方法 openFileOutput(String,int)的第 1 个参数为要读取的文件名，第 2 个参数为读取方式。

【例 4.1.3】　内部文件读/写示例。

在文本编辑框内输入信息后，单击"保存信息"按钮，即可以追加方式保存至文件 text 中。单击"读取信息"按钮，即可将文件内容以 Toast 方式显示。程序运行界面如图 4.1.6 所示。

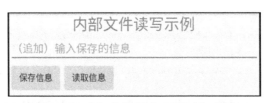

图 4.1.6　内部文件读/写程序运行界面

程序代码如下：

```java
public class MainActivity extends AppCompatActivity implements View.OnClickListener{
    EditText et;   //文本编辑框
    Button btn1,btn2;    //保存信息按钮、读取信息按钮
    @Override
    protected void onCreate(Bundle savedInstanceState) {
        super.onCreate(savedInstanceState);
        setContentView(R.layout.activity_main);
        et=findViewById(R.id.et);
        btn1=findViewById(R.id.btn1);btn1.setOnClickListener(this);
        btn2=findViewById(R.id.btn2);btn2.setOnClickListener(this);
    }
```

```java
@Override
public void onClick(View v) {
    int se=v.getId();
    switch (se){
        case R.id.btn1:
            String saveInfo=et.getText().toString();
            try {
                //第 1 个参数为文件名；第 2 个参数为 Context 定义的静态常量（Context.
                  可省略）
                FileOutputStream fos = openFileOutput("text", Context.MODE_
                                                                  APPEND);
                //追加方式写入
                fos.write(saveInfo.getBytes());
                fos.close();
                Toast.makeText(this, "文件保存成功！ ", Toast.LENGTH_SHORT).show();
            } catch (Exception e) {
                e.printStackTrace();
            }
            break;
        case R.id.btn2:   //读出、显示
            String fileContent="";
            try {
                FileInputStream fis = openFileInput("text");
                //定义大小为文件长度的字节数组
                byte[] buffer=new byte[fis.available()];
                //读文件内容到数组
                fis.read(buffer);
                //创建字节数组对应的字符串
                fileContent=new String(buffer);
                fis.close();
            } catch (Exception e) {
                e.printStackTrace();
            }
            Toast.makeText(this, fileContent, Toast.LENGTH_SHORT).show();
    }
}
```

注意：

（1）Android 内部存储，并不需要在清单文件里注册任何权限。

（2）文件 text 为应用程序私有，保存在 Android 设备的文件夹\data\data\包名\files 里。

3．Android 外部存储

Android 的不同版本，其外部存储的目录有差别。

2.3 版本之前存储在/sdcard 中，4.3 版本之前是在/mnt/sdcard 中，4.3 版本之后是在/storage/sdcard 中。在高版本（6.0）之后，为了兼容以前的版本，使用 sdcard 和 mnt/sdcard

这两个指向真实目录 storage/emulated/0 的符号链接。

外部存储不仅需要在清单文件里注册外部存储的读/写权限，还需要在 Activity 组件程序里动态申请权限。

4.1.4 抽象类 android.net.Uri 及其静态方法 parse()

类 java.net.URL 是用来描述 Internet 信息资源的。例如，表示 Web 服务器资源的字符串 http://www.wustwzx.com/as/index.html。又如，表示 FTP 服务器资源的字符串 ftp://39.109.11.33。

注意：统一资源定位器 URL 字符串里前面的 http 和 ftp 等，都是访问协议。

在 Android 开发中，使用比 URL 更加通用的概念 URI 来描述资源。URI 是通用资源标识符（Universal Resource Identifier）的英文缩写。

类 android.net.Uri 用来表示 URI，Android API 中许多方法参数是 Uri 类型，使用静态方法 Uri.parse()实现表示 URI 字符串到 Uri 对象的转化。例如，Uri.parse("tel:02751012663")的值就是 Uri 类型。

注意：

（1）Android API 中，有许多方法的参数类型为 Uri 类型。

（2）表示内容提供者路径的 Uri，以 "content://" 开头。

4.2 打电话程序设计

打电话是手机的基本功能之一，下面介绍两种实现方式。

1. 使用系统自带的拨号程序拨打电话

调用系统自带的拨号程序，不需要任何权限，其实现代码如下：

```
Intent intent=new Intent();
intent.setAction(Intent.ACTION_DIAL);   //拨号
intent.setData(Uri.parse("tel:15527643858"));   //意图数据，任选
/*
    //显示用户数据的通用方式，根据用户的数据类型打开相应的 Activity
    intent.setAction(Intent.ACTION_VIEW);
    intent.setData(Uri.parse("tel:15527643858"));*/
startActivity(intent);
```

程序运行时，进入系统的拨号界面。

2. 立即拨打指定电话

立即拨打电话，不会调用系统的拨号程序，在用户输入电话并单击按钮后立即呼叫，效果如图 4.2.1 所示。

在清单文件中，注册电话权限：

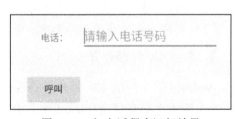

图 4.2.1 打电话程序运行效果

```xml
<uses-permission android:name="android.permission.CALL_PHONE" />
```

Activity 组件包含了危险权限的动态检测与申请,其完整代码如下:

```java
/*
    立即拨打指定的电话,不同于使用系统的拨号程序
    需要配置权限 android.permission.CALL_PHONE
*/
public class MainActivity extends AppCompatActivity {
    EditText editText;
    @Override
    protected void onCreate(Bundle savedInstanceState) {
        super.onCreate(savedInstanceState);
        setContentView(R.layout.activity_main);
        if (ActivityCompat.checkSelfPermission(MainActivity.this, Manifest.permission.
                        CALL_PHONE) != PackageManager.PERMISSION_GRANTED) {
            //没有权限时请求该权限:出现"是否型"对话框由用户选择是否授权
            ActivityCompat.requestPermissions(MainActivity.this, new String[]{
                        Manifest.permission.CALL_PHONE}, 1);
        }else{
            fun();   //执行有权限时的业务逻辑,fun()需要用户定义
        }
    }
    void  fun(){
        editText=findViewById(R.id.editText);
        findViewById(R.id.button).setOnClickListener(new View.OnClickListener() {
            @Override
            public void onClick(View v) {
                Intent intent=new Intent();
                intent.setAction(Intent.ACTION_CALL);    //意图动作
                intent.setData(Uri.parse("tel:"+editText.getText()));   //意图数据
                startActivity(intent);
            }
        });
    }
    @Override
    //Android 6.0 动态权限处理的回调方法
    public void onRequestPermissionsResult(int requestCode, @NonNull String[] permissions,
                                            @NonNull int[] grantResults) {
        super.onRequestPermissionsResult(requestCode, permissions, grantResults);
        switch (requestCode) {
            case 1:
                //所需关键权限
                if (grantResults[0] == PackageManager.PERMISSION_GRANTED) {
                    fun();
                } else {
                    Toast.makeText(this, "没有打电话权限", Toast.LENGTH_LONG). show();
```

```
                    finish();
                }
            }
        }
    }
```

注意：拨打固定电话时，需要在电话前加区号。

4.3 短信程序设计

4.3.1 SMS 简介

SMS（Short Message Service，短信息服务）是一种存储和转发服务。短信息并不是直接从发信人发送到接收人，而是通过 SMS 中心进行转发的。如果接收人处于未连接状态（手机没有信号或者关机），则短信息于接收人再次连接时发送。

4.3.2 短信管理器

短信管理器类 android.telephony.SmsManager 提供了静态方法 getDefault()来获取 SmsManager 实例对象，也提供了对短信文本进行分块（可能内容较多）的方法和短信发送的方法，如图 4.3.1 所示。

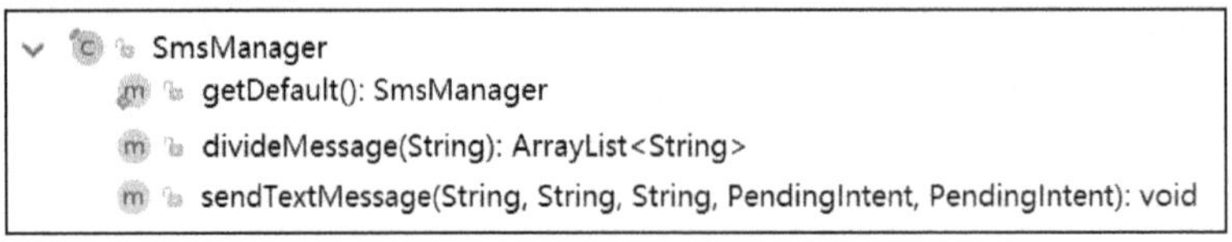

```
▼ ⓒ SmsManager
    ⓜ ⓑ getDefault(): SmsManager
    ⓜ ⓑ divideMessage(String): ArrayList<String>
    ⓜ ⓑ sendTextMessage(String, String, String, PendingIntent, PendingIntent): void
```

图 4.3.1　SmsManager 类及其常用方法

注意：方法 sendTextMessage()的 5 个参数中，一般只需填写第 1 个参数（手机号码）和第 3 个参数（短信内容），其他填写 null 即可。

4.3.3 短信发送程序的实现

1．使用系统的短信程序

调用系统自带的短信程序，不需要任何权限，其实现代码如下：

```
Intent intent=new Intent();
intent.setAction(Intent.ACTION_SENDTO); //意图动作
//intent.setAction(Intent.ACTION_VIEW);
intent.setData(Uri.parse("sms:15527643858?body=手机短信测试"));   //意图数据=>发短信
//另一种写法
/*Uri uri=Uri.parse("smsto:15527643858 ");
Intent intent=new Intent(Intent.ACTION_SENDTO,uri);
intent.putExtra("sms_body", "手机短信测试"); */
startActivity(intent);
```

程序运行时，进入系统的短信编辑界面，显示默认的短信文本。

2. 使用短信管理器，立即发送短信

使用短信管理器发送短信，需要在清单文件中注册如下两个权限：

```
<uses-permission android:name="android.permission.SEND_SMS"/>
<uses-permission android:name="android.permission. READ_PHONE_STATE "/>
```

【例 4.3.1】 使用短信管理器发送短信。

在主 Activity 中分别放置了用于输入短信内容的 TextView 控件和 EditView 控件，最后是一个 Button 命令按钮。用户输入电话号码和短信内容后单击按钮就会立即发送短信。Activity 组件程序代码如下：

```java
/*
    通过短信管理器发送短信，涉及两个危险权限组：
        （1）短信权限 android.permission.SEND_SMS
        （2）电话权限：android.permission.READ_PHONE_STATE
    小米手机在每次发送短信时，都会动态请求该权限，这与其他手机不同
*/
public class MainActivity extends AppCompatActivity {
    //定义危险权限数组
    static String[] permissions = {"android.permission.SEND_SMS",
                                    "android.permission.READ_ PHONE_STATE"};
    @Override
    protected void onCreate(Bundle savedInstanceState) {
        super.onCreate(savedInstanceState);
        setContentView(R.layout.activity_main);
        List<String> mPermissionList = new ArrayList<>();   //拟筛选未授权的权限
        mPermissionList.clear();
        for (int i = 0; i < permissions.length; i++) {
            if (ContextCompat.checkSelfPermission(this, permissions[i]) !=
                                            PackageManager.PERMISSION_GRANTED) {
                mPermissionList.add(permissions[i]);
            }
        }
        if (mPermissionList.isEmpty()) {
            fun();   //应用已经获得权限时直接执行
        } else{
            //集中请求未授权的权限；得到 List 对象对应的数组对象
            String[] needPermissions = mPermissionList.toArray(new String[mPermissionList.size()]);
            //请求权限后执行回调方法
            ActivityCompat.requestPermissions(this, needPermissions, 1);
        }
    }
    void  fun(){
        findViewById(R.id.btn_send).setOnClickListener(new View.OnClickListener() {
            @Override
```

```java
            public void onClick(View v) {
                EditText et_phone=findViewById(R.id.et_phone);
                EditText et_content=findViewById(R.id.et_content);
                //获取短信管理器
                SmsManager smsManager=SmsManager.getDefault();
                //短信内容分段
                List<String> list=smsManager.divideMessage(et_content.getText().toString());
                for (String sms:list){      //分段发送
                    //方法参数 1 为手机号，参数 3 为短信内容，共 5 个参数
                    smsManager.sendTextMessage(et_phone.getText().toString(),null,sms,null,null);
                }
            }
        });
    }
    @Override   //Android 6.0 动态权限处理的回调方法
    public void onRequestPermissionsResult(int requestCode, @NonNull String[] permissions,
                                           @NonNull int[] grantResults) {
        super.onRequestPermissionsResult(requestCode, permissions, grantResults);
        switch (requestCode) {
            case 1:    //危险权限组是 1 次请求，只有 1 个请求码
                for (int i = 0; i < grantResults.length; i++) {
                    if (grantResults[i] != PackageManager.PERMISSION_GRANTED) {
                        Toast.makeText(this, "权限不足，无法实现全部功能！",
                                Toast.LENGTH_SHORT).show();
                        finish();
                    }
                }
                fun();    //全部授权时执行
        }
    }
}
```

4.4 手机音频播放与录音程序设计

4.4.1 音频播放

媒体播放类 MediaPlayer 位于 android.media 包内，除了提供创建媒体播放机的构造方法外，还提供创建媒体播放机的静态方法 create()。当然，播放方法 start()和停止方法 stop()是肯定会提供的。媒体播放类 MediaPlayer 的定义如图 4.4.1 所示。

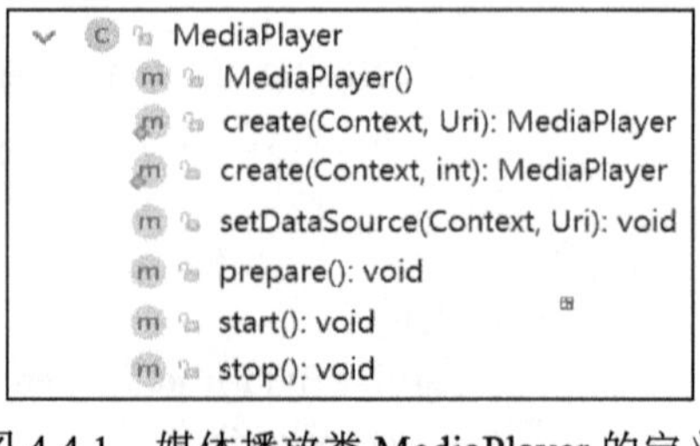

注意：用 MediaPlayer 播放项目自带的音乐

图 4.4.1 媒体播放类 MediaPlayer 的定义

文件时，使用静态方法 create()创建其实例对象；播放外部音乐文件时，则需要先使用构造方法创建 MediaPlayer 实例对象。

【例 4.4.1】 播放外部存储的音频文件。

运行程序时，播放手机外部存储区的音乐文件/music/white.mp3，按返回键将关闭应用窗口，同时停止音乐播放。在项目的清单文件里，需要注册如下权限：

```
<uses-permission android:name="android.permission. READ_EXTERNAL_STORAGE "/>
```

Activity 组件程序的代码如下：

```
/*
    按返回键将执行 Activity 的生命周期方法 onDestrory()，其内包含停止音乐播放的代码
*/
public class MainActivity extends AppCompatActivity {
    MediaPlayer mp;
    @Override
    protected void onCreate(Bundle savedInstanceState) {
        super.onCreate(savedInstanceState);
        setContentView(R.layout.activity_main);
        if (ActivityCompat.checkSelfPermission(MainActivity.this, Manifest.permission.
            READ_EXTERNAL_STORAGE)!= PackageManager.PERMISSION_GRANTED) {
            //没有权限时请求该权限：出现"是否型"对话框由用户选择是否授权
            ActivityCompat.requestPermissions(MainActivity.this, new String[]{
                            Manifest.permission.READ_EXTERNAL_STORAGE}, 1);
        }else{
            fun();   //执行有权限时的业务逻辑，fun()需要用户定义
        }
    }
    //已经获得权限（对应已经安装并授权）或者动态授权（对应于首次安装）后执行的代码块
    public void fun() {
        //android studio 获取外部存储根路径
        String basePath = Environment.getExternalStorageDirectory().getPath();
        //特定目录里的文件
        String filePath = basePath + "/music/white.mp3";
        Log.i("wutest", filePath);
        MediaPlayer mp = new MediaPlayer();
        try {
            mp.setDataSource(filePath);   //需要权限 READ_EXTERNAL_STORAGE
            mp.prepare();
            mp.start();
        } catch (IOException e) {
            //打开日志信息，在监视窗口中可过滤出来显示
            Log.i("wutest", "file path errror or no permission of READ_EXTERNAL_STORAGE");
            e.printStackTrace();
        }
    }
```

```
@Override    //Android 6.0 动态权限处理的回调方法
public void onRequestPermissionsResult(int requestCode, @NonNull String[] permissions,
                                              @NonNull int[] grantResults) {
    super.onRequestPermissionsResult(requestCode, permissions, grantResults);
    switch (requestCode) {
        case 1.
            if (grantResults[0] == PackageManager.PERMISSION_GRANTED) {
                fun();
            } else {
                Toast.makeText(this, "没有读 SD 卡权限", Toast.LENGTH_LONG).show();
                finish();
            }
    }
}
@Override
protected void onDestroy() {    //按返回键将执行此方法
    super.onDestroy();
    mp.stop();
    mp=null;
}
}
```

注意：为了能访问 Android 10 及以上系统外部存储的照片和音乐，需要在项目清单文件的标签<application>里添加一个属性：android:requestLegacyExternalStorage="true"。

4.4.2 手机录音

Android 提供了媒体播放类 MediaPlayer，也提供了媒体录音类 MediaRecorder。类 MediaRecorder 提供的常用方法，如图 4.4.2 所示。

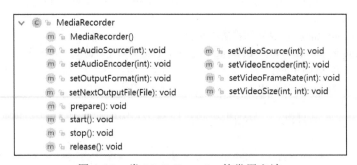

图 4.4.2 类 MediaRecorder 的常用方法

【例 4.4.2】 手机录音程序设计。

运行程序时，在主窗口中出现"开始录音""停止录音"两个按钮，单击"开始录音"按钮后开始录音，单击"停止录音"按钮后停止录音。其中，录音文件保存在 SD 卡的 audioRecords 文件夹里，文件名为 AudioRecord×××.amr。其中，"×××"为若干位随机数字。Activity 组件程序代码如下：

```java
/*
    手机录音程序，需要危险权限：录音权限和外部存储权限
    录音文件保存在文件夹 audioRecords 里
*/
public class MainActivity extends AppCompatActivity implements View.OnClickListener{
    ImageButton st,stop;
    MediaRecorder mRecorder;
    File recordPath;
    File recordFile;
    //定义危险权限数组
    static String[] permissions = {"android.permission.RECORD_AUDIO",
                                    "android.permission.WRITE_EXTERNAL_STORAGE"};
    @Override
    public void onCreate(Bundle savedInstanceState) {
        super.onCreate(savedInstanceState);
        setContentView(R.layout.activity_main);
        List<String> mPermissionList = new ArrayList<>();    //拟筛选未授权的权限
        mPermissionList.clear();
        for (int i = 0; i < permissions.length; i++) {
            if (ContextCompat.checkSelfPermission(this, permissions[i]) !=
                                        PackageManager.PERMISSION_GRANTED) {
                mPermissionList.add(permissions[i]);
            }
        }
        if (mPermissionList.isEmpty()) {
            fun();   //应用已经获得权限时直接执行
        } else{
            //集中请求未授权的权限；得到 List 对象对应的数组对象
            String[] needPermissions = mPermissionList.toArray(new String[mPermissionList.
                                                                          size()]);
            //请求权限后执行回调方法
            ActivityCompat.requestPermissions(this, needPermissions, 1);
        }
    }
    public void fun(){    //有全部权限时的业务逻辑
        st = findViewById(R.id.st);
        stop = findViewById(R.id.stop);
        st.setOnClickListener(this);
        stop.setOnClickListener(this);
    }
    @Override
    public void onClick(View v) {    //因为主 Activity 实现了接口 OnClickListener
        if(v == st){
            startRecord();
            Toast.makeText(getApplicationContext(), "开始录音......", Toast.LENGTH_LONG).
                                                                              show();
```

```java
        }
        if(v == stop){
            stopRecord();
            Toast.makeText(getApplicationContext(), "录音结束!", Toast.LENGTH_SHORT).
                    show();
        }
    }
    public void startRecord() {
        if (checkSDCard()) {
            //设置录音文件的路径及文件名
            recordPath = Environment.getExternalStorageDirectory();
            File path = new File(recordPath.getPath() + File.separator + "audioRecords");
            //创建子目录
            if(!path.mkdirs()){
                Toast.makeText(this, "目录创建失败",Toast.LENGTH_LONG);
            }
            recordPath=path;
        } else {
            Toast.makeText(getApplicationContext(), "SD Card 未连接", Toast.LENGTH_LONG).
                    show();
            return ;
        }
        try {
            //使用静态方法 File.createTempFile()建立的文件名后多了几个随机的数字
            recordFile = File.createTempFile(String.valueOf("MyAudioRecord"), ".amr",
                    recordPath);
        } catch (IOException e) {
            Toast.makeText(this, "文件创建失败",Toast.LENGTH_LONG).show();
        }
        mRecorder = new MediaRecorder();
        mRecorder.setAudioSource(MediaRecorder.AudioSource.MIC);        //设置麦克风
        mRecorder.setOutputFormat(MediaRecorder.OutputFormat.DEFAULT);  //文件格式
        mRecorder.setAudioEncoder(MediaRecorder.AudioEncoder.DEFAULT);  //音频编码
        mRecorder.setOutputFile(recordFile.getAbsolutePath());          //文件路径
        try {
            mRecorder.prepare();
            mRecorder.start();      //开始录音
        } catch (IllegalStateException e) {
            e.printStackTrace();
        } catch (IOException e) {
            e.printStackTrace();
        }
    }
    public void stopRecord() {
        try{
            if(mRecorder != null) {
```

```
                mRecorder.stop();
                mRecorder.release();
                mRecorder = null;
            }
        }
        catch(IllegalStateException e){
            e.printStackTrace();
        }
    }
    private boolean checkSDCard() {    //检查是否存在SD卡
        if(android.os.Environment.getExternalStorageState().
                                    equals(android.os.Environment.MEDIA_MOUNTED))
            return true;
        return false;
    }
    @Override    //Android 6.0 动态权限处理的接口回调方法
    public void onRequestPermissionsResult(int requestCode, @NonNull String[] permissions,
                                            @NonNull int[] grantResults) {
        super.onRequestPermissionsResult(requestCode, permissions, grantResults);
        switch (requestCode) {
            case 1:
                for (int i = 0; i < grantResults.length; i++) {
                    if (grantResults[i] != PackageManager.PERMISSION_GRANTED) {
                        Toast.makeText(this, "权限不足,无法实现全部功能!",
                                            Toast.LENGTH_SHORT).show();
                        finish();
                    }
                }
                fun();    //全部授权时执行
        }
    }
}
```

4.5 手机视频播放

控件 MediaPlayer 不仅能播放音频,还可以播放视频。尽管 Android 也提供了专门播放视频的控件 VideoView,其实是调用 MediaPlayer 来播放视频的,并提供了一些视频播放的辅助功能。VideoView 位于软件包 android.widget 内,其常用方法如图 4.4.3 所示。

注意:

(1) MediaPlayer 与 VideoView 不是处于同一软件包内。

(2) VideoView 播放完成监听器接口方法的参数为 MediaPlayer 类型。

作为协作类的 MediaController 与控件 VideoView 配套使用,播放时提供了暂停、向前和向后的控制功能。

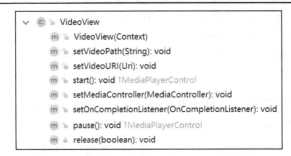

图 4.4.3　视频播放控件 VideoView 的常用方法

播放外部存储的视频文件，当然需要读外部存储文件的权限：

`<uses-permission android:name="android.permission.READ_EXTERNAL_STORAGE"/>`

Activity 组件程序里的主要实现代码如下：

```
String basePath = Environment.getExternalStorageDirectory().getPath();
//特定目录里的视频文件
String filePath = basePath + "/Movies/test.mp4";
VideoView videoView = findViewById(R.id.videoView);
try {
    videoView.setVideoURI(Uri.parse(filePath));
    MediaController mediaController = new MediaController(this);
    videoView.setMediaController(mediaController);
    videoView.start();
} catch (Exception e) {
    e.printStackTrace();
}
videoView.setOnCompletionListener(new MediaPlayer.OnCompletionListener() {
    @Override
    public void onCompletion(MediaPlayer arg0) {
        finish();   //播放完毕后关闭当前的 Activity
    }
});
```

4.6　手机拍照程序设计

目前的 Android 手机都内置了摄像头，为拍照提供了硬件支持，它也提供了拍照程序。下面通过两个案例分别说明使用自带的手机拍照程序和使用摄像头 API 实现拍照的方法。

【例 4.6.1】　使用手机自带的相机程序拍照，并将结果写入 SD 卡文件。

程序运行后，调用系统自带的相机程序，出现取景界面。按下相机的快门后，出现是否保存的对话框，如图 4.6.1 所示。

第4章 手机基本功能程序设计

图 4.6.1 拍照后询问是否保存对话框

单击✓按钮（有的手机出现的是文字按钮）后，文件将自动保存至 SD 卡的 myImage 文件夹里，返回至主 Activity，在 ImageView 控件里显示图片。

本应用使用手机自带的相机程序，因此，不需要相机权限，只需要外部存储权限，在清单文件里注册的代码如下：

```
<uses-permission android:name="android.permission.WRITE_EXTERNAL_STORAGE"/>
```

Activity 组件程序代码如下：

```
public class MainActivity extends AppCompatActivity {
    ImageView pic;
    @Override
    protected void onCreate(Bundle savedInstanceState) {
        super.onCreate(savedInstanceState);
        setContentView(R.layout.activity_main);
        if (ActivityCompat.checkSelfPermission(MainActivity.this,Manifest.permission.
            WRITE_EXTERNAL_STORAGE) != PackageManager.PERMISSION_GRANTED) {
            //没有权限时请求该权限：出现"是否型"对话框由用户选择是否授权
            ActivityCompat.requestPermissions(MainActivity.this, new String[]
                            {Manifest.permission.WRITE_EXTERNAL_STORAGE}, 1);
        } else {
            fun();
        }
    }
    void fun(){
        pic=findViewById(R.id.pic);
        Intent intent=new Intent(MediaStore.ACTION_IMAGE_CAPTURE); //调用系统相机程序
        startActivityForResult(intent,1);
    }
    @Override
```

```java
protected void onActivityResult(int requestCode, int resultCode, Intent data) {
    super.onActivityResult(requestCode, resultCode, data);
    String basePath,filePath;                //外部存储根路径、文件保存路径
    if(Build.VERSION.SDK_INT<29){            //Android 10 对应的 API 级别为 29
        // 指向外部存储根路径：(1) 模拟器 /sdcard      (2) 手机 /stroage/sdcard0
        basePath = Environment.getExternalStorageDirectory().getPath();
    } else{  //Android 10 之后，不能直接在根路径下创建子文件夹
        //华为手机指向的外部存储路径：/stroage/sdcard0/Android/data/应用的包名/files
        basePath = this.getExternalFilesDir(null).getPath();
    }
    filePath = basePath + "/myImage";
    if(requestCode==1){
        if(resultCode==RESULT_OK){           //按确定时
            Bundle bundle=data.getExtras();
            Bitmap bitmap=(Bitmap) bundle.get("data");
            FileOutputStream fos=null;
            File file=new File(filePath);
            if(!file.exists()) file.mkdir();     //创建文件夹
            String fileName=filePath+"/111.jpg";
            try {
                fos=new FileOutputStream(fileName);
                bitmap.compress(Bitmap.CompressFormat.JPEG,100,fos);
            }catch (Exception e){
                e.printStackTrace();
            }finally {
                try {
                    fos.flush();
                } catch (IOException e) {
                    e.printStackTrace();
                }
            }
            pic.setImageBitmap(bitmap);
        }
    }
}
@Override   //Android 6.0 动态权限处理的接口回调方法
public void onRequestPermissionsResult(int requestCode, @NonNull String[] permissions,
                                       @NonNull int[] grantResults) {
    super.onRequestPermissionsResult(requestCode, permissions, grantResults);
    switch (requestCode) {
        case 1:
            if (grantResults[0] == PackageManager.PERMISSION_GRANTED) {    //所需关
                                                                           //键权限
                fun();
            } else {
                Toast.makeText(this, "没有写 SD 卡权限", Toast.LENGTH_LONG).show();
                finish();
            }
    }
}
```

习 题 4

一、判断题

1．Activity 组件调用时，使用 Intent 携带的数据类型必须是 Bundle 类型。
2．内部文件与外部文件的读/写，都需要在清单文件里注册读/写权限。
3．在 Activity 组件里请求危险权限时，为了让确认的权限立即生效，必须重写回调方法 onRequestPermissionsResult()。
4．首次安装、运行 Android 6.0 应用时，动态权限请求弹窗的次数与该应用涉及的危险权限组的个数相同。
5．MediaPlayer 和 VideoView 是 Android 提供的两个独立无关的视频播放控件。

二、选择题

1．假定使用方法 startActivityForResult()进行组件调用。当从子 Activity 组件返回主调 Activity 组件时，将执行主调组件里的____方法。
 A．onDestrory()　　　　　　　　B．onActivityResult()
 C．onRestart()　　　　　　　　　D．onBind()
2．下列 Android 权限中，不需要在 Activity 组件程序里动态申请的是____。
 A．Internet　　　B．SMS　　　C．Location　　　D．Storage
3．下列选项中，____不是类 android.telephony. SmsManager 提供的方法。
 A．getDefault()　　　　　　　　B．handleMessage
 C．sendTextMessage()　　　　　D．divideMessage()
4．使用短信管理器发送短信，除了需要注册 SEND_SMS 权限外，还需要的是____。
 A．READ_SMS　　　　　　　　B．CALL_PHONE
 C．READ_PHONE_STATE　　　D．READ_CALL_LOG
5．Activity 组件的回调方法 onActivityResult()包含的参数个数是____。
 A．1　　　B．2　　　C．3　　　D．4

三、填空题

1．获取 Activity 组件调用时传递的捆绑数据，应使用 Intent 对象的____方法。
2．播放 SD 卡里的音乐文件，需要在清单文件里注册____权限。
3．为了获取特定的 Activity 组件调用返回时携带的数据，需要使用____码。
4．编写手机拍照程序时，为了取得手机内置的相机程序返回的数据，需要使用____方法去启动手机自带的相机程序。
5．检查是否存在 SD 卡、获取外在存储文件的根路径等，均需要使用类 android.os.____。

实 验 4

一、实验目的

（1）掌握带数据传递的 Activity 组件调用程序设计。
（2）掌握 Android 6.0 的动态权限模型。
（3）掌握 Android 共享存储与文件存储的用法。
（4）掌握打电话程序的设计方法。
（5）掌握发送短信程序的设计方法。
（6）掌握手机音频播放与录音程序的设计方法。
（7）掌握手机视频播放程序的设计方法。

二、实验内容及步骤

【预备】在 Android Studio 中，新建名为 Example4 的项目后，访问本课程配套的网站 http://www.wustwzx.com/as/sy/sy04.html，复制相关代码，完成如下几个模块的设计。

1．有数据传递的 Activity 组件调用程序设计（参见例 4.1.1）

（1）在项目里，新建名为 example4_1 的模块，在布局文件里添加 2 个 Button 控件。

（2）使用菜单 File→New→Activity，创建名为 SecondActivity 的组件。在清单文件里，使用标签<intent-filter>配置对这个 Activity 组件的隐式调用，再创建名为 ThirdActivity 的组件，但不配置隐式调用代码。组件 SecondActivity 和 ThirdActivity 都有自己的布局。

（3）在 Activity 组件程序的 onCreate()方法里，分别定义两个按钮的单击事件处理代码，主要包括使用方法 startActivityForResult()隐式调用和显式调用，使用请求码、通过 Intent 对象携带数据进行调用，在回调方法 onActivityResult()里通过结果码获取指定组件返回的数据、通过请求码判断从哪个组件返回。

（4）在被调组件里，使用方法 getIntent()获取主调组件的意图对象，进而获取数据，使用方法 setResult()设置结果码。

（5）部署本模块并做运行测试。

2．使用接口 SharedPreferences 实现的共享存储程序设计（参见例 4.1.2）

（1）在项目里，新建名为 example4_2 的模块，在布局文件里添加 2 个 EditText 控件、2 个 Button 控件和 1 个 CheckBox 控件。

（2）在 Activity 组件程序的 onCretae()方法里，获取接口 SharedPreferences 类型的对象，编写登录按钮的单击监听器代码。

（3）对 CheckBox 控件对象应用选中监听器方法 setOnCheckedChangeListener()，使用内部接口 SharedPreferences.Editor 的实例对象，保存用户的登录信息。

(4)部署本模块并做运行测试。

3．Android 内部文件读/写程序设计（参见例 4.1.3）

(1)在项目里，新建名为 example4_3 的模块，在布局文件里添加 1 个 EditText 控件和 2 个 Button 控件。

(2)在保存信息按钮的单击事件代码里，使用方法 openFileOutput()创建文件输出 FileOutputStream 对象，进而完成文件的写入。

(3)在获取信息按钮的单击事件代码里，使用方法 openFileInput()创建 FileInputStream 对象，进而完成文件的读出。

(4)部署本模块并做运行测试。

4．电话程序设计（参见第 4.2 节）

(1)在项目里，新建名为 example4_4 的模块。
(2)在 Activity 组件程序的 onCreate()方法里，编写使用系统自带的拨号程序代码。
(3)部署本模块并做运行测试。
(4)在清单文件里，注册电话权限 android.permission.CALL_PHONE。
(5)在布局文件里，添加 1 个 EditText 控件和 1 个 Button 控件。
(6)在 Activity 组件程序，动态申请权限（涉及 1 个权限组）、编写按钮的单击监听器代码，使用意图 Intent.ACTION_CALL 完成电话的立即拨打。
(7)再次部署本模块，并做运行测试。

5．短信程序设计（参见第 4.3 节）

(1)在项目里，新建名为 example4_5 的模块。
(2)使用系统自带的短信程序，完成短信的发送。
(3)部署本模块并做运行测试。
(4)在清单文件里，注册发送短信权限 android.permission.SEND_SMS 和读取电话状态权限 android.permission.READ_PHONE_STATE。
(5)在布局文件里，添加 1 个 EditText 控件和 1 个 Button 控件。
(6)在 Activity 组件程序里，动态申请权限（涉及两个权限组）、编写按钮的单击监听器代码，使用短信管理器完成短信的立即发送。
(7)再次部署本模块，并做运行测试。

6．手机拍照程序设计（参见例 4.6.1）

(1)在项目里，新建名为 example4_6 的模块，在布局文件里添加 1 个 ImageView 控件对象。

(2)在清单文件里，注册权限 android.permission.WRITE_EXTERNAL_STORAGE。
(3)在 Activity 组件程序里，动态申请权限（涉及 1 个权限组）、使用有返回结果的方法 startActivityForResult()调用系统的相机程序，在回调方法 onActivityResult()里编写保存图像的代码。最后，刷新图像控件对象。

（4）部署本模块并做运行测试。

三、实验小结及思考

（由学生填写，重点写上机中遇到的问题。）

第 5 章 服务组件及其应用

服务组件 Service 也是 Android 的重要组件,它表示运行于后台且没有界面的服务程序。与 Activity 组件一样,Service 也需要在清单文件里注册、由 Intent 对象激活。本章学习要点如下:
- 服务组件 Service 的特点及生命周期;
- 服务的多种启动及停止方式;
- 系统提供的常用服务;
- 掌握远程服务的使用(含 AIDL)。

5.1 服务组件 Service 的基本用法

5.1.1 Android 系统服务

Android 后台运行的很多服务组件 Service 是在系统启动时被开启的,以支持系统的正常工作。比如,MountService 监听是否有 SD 卡安装及移除,ClipboardService 提供剪切板功能,PackageManagerService 提供软件包的安装、移除及查看等。Android 提供的常用服务,如表 5.1.1 所示。

5.1.1 Android 中常用的系统服务

服 务 名 称	功 能 描 述
NOTIFICATION_SERVICE	通知服务,得到通知管理器对象
POWER_SERVICE	电源服务,得到电源管理器对象
AUDIO_SERVICE	实现对音量、音效、声道及铃声等的管理服务
WIFI_SERVICE	WiFi 服务,得到 WiFi 管理器对象
BLUETOOTH_SERVICE	蓝牙服务
LOCATION_SERVICE	调用此定位服务,获得位置管理器对象,实现对 GPS 的使用
CONNECTIVITY_SERVICE	调用网络连接(GPRS 或 WiFi)服务,获得 ConnectivityManager 对象

下面对 Android 提供的若干系统服务及其相关类加以说明。

1．应用程序包管理服务 PackageManagerService

通过 PackageManagerService（PMS）获取已经安装应用程序的相关信息，根据应用的包名可启动该应用。

2．活动管理服务 ActivityManagerService 及其内部类

调用系统服务 ActivityManagerService（AMS）得到活动管理器 ActivityManager 对象。活动管理器类 ActivityManager 的内部类 RunningServiceInfo 与服务相关，它封装了正在运行的服务程序的信息。

3．窗口管理服务 WindowManagerService

窗口管理服务 WindowManagerService（WMS）是 Android Framework 的核心服务，用于对窗口管理。

4．通知服务 NOTIFICATION_SERVICE

调用系统服务 NOTIFICATION_SERVICE，得到 NotificationManager 对象，进而管理状态栏通知。

5．网络管理服务 NetworkManagementService

Android 系统网络连接和管理服务，提供了 ConnectivityService 等四个服务。

6．WiFi 服务 WiFi Service

调用系统服务 WiFi Service，得到 WiFiManager 对象，进而管理 WiFi 信息。

7．蓝牙管理服务类 BluetoothManagerService

BluetoothManagerService 负责蓝牙后台管理和服务。

8．位置管理器服务 LocationManagerService

LocationManagerService 提供了位置服务、GPS 服务和定位服务等。

9．内容服务 ContentService

ContentService 主要为数据库等提供解决方法的服务。

10．剪贴板服务 ClipboardService

ClipboardService 提供了剪贴板服务。

11．闹铃和定时器类 AlarmManagerService

AlarmManagerService 提供了闹铃和定时器等功能。

12．音频服务类 AudioService

AudioService 提供了音量、音效、声道及铃声等的管理服务。

13．电源管理服务 PowerManagerService

PowerManagerService 提供电源管理服务。

14．电池管理类 BatteryService

电池管理类 BatteryService 负责监控电池的充电状态、电池电量、电压、温度等信息，当电池信息发生变化时，发出广播通知其他关系电池信息的进程和服务。

15．屏保服务 DreamManagerService

DreamManagerService 提供屏幕保护功能。

16．输入法服务 InputMethodManagerService

InputMethodManagerService 提供打开和关闭输入法的功能。

17．光感应传感服务 LightsService

LightsService 提供光感应传感器服务。

18．锁屏设置服务 LockSettingsService

LockSettingsService 提供锁屏、手势等安全服务。

19．壁纸管理服务 WallpaperManagerService

WallpaperManagerService 提供壁纸管理服务。

20．搜索服务 SearchManagerService

SearchManagerService 提供搜索服务。

21．状态栏管理服务 StatusBarManagerService

StatusBarManagerService 提供对状态栏的管理功能。

22．振动服务 VibratorService

VibratorService 提供振动器服务。

调用系统服务方法 getSystemService()，以类 Context 中定义的系统服务常量为参数，其返回值为 Object 类型。因此，在实际应用时，需要进行类型强制转换（为相应的管理器对象类型）。

5.1.2 Service 组件及其生命周期

Service 可以在后台执行很多任务，如处理网络事务、播放音乐等。它是 Android 系统的服务组件，适用于开发没有用户界面且长时间在后台运行的功能。Service 有利于降低系统资源的开销，而且比 Activity 有更高的优先级，Service 不会在系统资源紧张时被 Android 系统优先终止。

在包 android.app 内，提供抽象类 Service，其定义如图 5.1.1 所示。

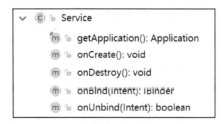

图 5.1.1　抽象类 Service 定义

启动和停止 Service 通常是由其他组件完成的。例如，组件 Activity 的超类 Context 提供启动和停止 Service 的方法 startService()和 stopService()等，如图 5.1.2 所示。

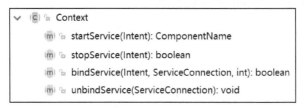

图 5.1.2　Context 提供关于 Service 的启动与停止方法

方法 bindService()是启动服务的另一个方式，称为绑定服务。相应地，使用方法 unbindService()取消绑定服务。Service 组件的生命周期，如图 5.1.3 所示。

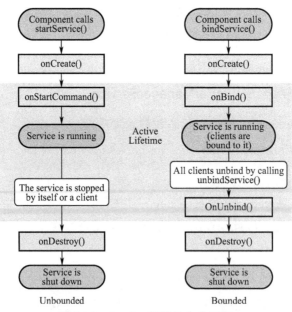

图 5.1.3　Service 组件的生命周期

注意：如同 Activity 组件一样，Service 组件必须在清单文件里使用标签<service>注册，否则，服务无法启动。使用 Android Studio 菜单 File→New→Service 创建的服务组件，将自动在清单文件里注册。抽象类 Service 也提供了停止服务的方法。

5.1.3 自定义服务与服务注册

自定义（新建）一个服务程序 Service，需要继承抽象类 android.app.Service，根据选用的服务启动方式，相应地重写 Service 的 onCreate()、onBinder()和 onDestroy()等生命周期方法。

自定义服务需要在清单文件里使用<service>标签注册，类似于 Activity 组件需要在清单文件中使用<activity>标签注册。

5.1.4 服务的显式启动与隐式启动

启动服务需要创建 Intent 对象。使用非绑定方式，在 Intent 对象里设置 Service 所在的类，并调用方法 startService(Intent)启动 Service，这种方式称为显式启动。

隐式启动是指在注册 Service 的同时，还内嵌了标签<intent-filter>及<action>。其中，<action>标签的 android:name 属性指定了引用该服务的名称。

隐式启动的好处是不需要指明需要启动哪一个 Activity，而由 Android 系统根据 Intent 的动作和数据来决定，这样有利于降低 Android 组件之间的耦合度，强调 Android 组件的可复用性。

注意：若 Service 与调用它的组件在同一个应用程序中，则既可以使用显式启动，也可以使用隐式启动（显式启动代码更简洁）；若 Service 和调用服务的组件在不同的应用程序中，则只能使用隐式启动方式。

【例 5.1.1】 调用后台音频播放服务。

程序运行时，显示两个分别用于播放与停止后台音乐的按钮。主要设计步骤如下：

（1）新建名为 Example5 的项目，再新建名为 examle5_1 的模块。

（2）在模块 examle5_1 的程序包名里，新建名为 MyAudioService 的 Service 组件，其代码如下：

```
/*
    Service 是编写服务组件的抽象基类
    onCreate()和 onDestroy()是 Service 的 2 个重要生命周期方法，需要在其内写代码
    非绑定方式启动服务时，不需要在生命周期方法 onBind()内写代码
*/
public class MyAudioService extends Service {
    MediaPlayer mp;
    @Override
    public void onCreate() {                //在开始服务时调用
        super.onCreate();
        mp = MediaPlayer.create(this,R.raw.white);
        //mp = MediaPlayer.create(getApplicationContext(), R.raw.white);    //也 OK
        mp.start();
    }
    @Override
    public void onDestroy() {               //在停止服务时调用
```

```java
        super.onDestroy();
        mp.stop();
        if(mp!=null) mp=null;
    }
    @Override
    public IBinder onBind(Intent intent) {    //不可省略的生命周期方法
        // TODO: Return the communication channel to the service.
        throw new UnsupportedOperationException("Not yet implemented");
    }
}
```

（3）编写 MainActivity 程序，调用服务组件，其代码如下：

```java
public class MainActivity extends AppCompatActivity implements View.OnClickListener{
    Intent intent;
    Button btn_play,btn_stop;
    @Override
    protected void onCreate(Bundle savedInstanceState) {
        super.onCreate(savedInstanceState);
        setContentView(R.layout.activity_main);
        btn_play = findViewById(R.id.btn_play);btn_play.setOnClickListener(this);
        btn_stop=findViewById(R.id.btn_stop);btn_stop.setOnClickListener(this);
    }
    @Override
    public void onClick(View v) {
        intent=new Intent(this,MyAudioService.class);        //显式调用服务意图
        /* //隐式调用时使用，需要意图过滤器
        intent=new Intent("com.example.audio_service_test.MAS");
        intent.setPackage("com.example.example5_1");    //隐式调用时还需要指定应用的包名*/
        int id=v.getId();
        switch (id){
            case R.id.btn_play:
                startService(intent);
                Toast.makeText(this, "音乐播放服务进行中...", Toast.LENGTH_SHORT). show();
                btn_stop.setEnabled(true);btn_play.setEnabled(false);
                break;
            case R.id.btn_stop:
                stopService(intent);
                btn_stop.setEnabled(false);btn_play.setEnabled(true);
        }
    }
    @Override
    protected void onDestroy() {                        //考虑播放时按返回键
        super.onDestroy();
        if(intent!=null) stopService(intent);           //停止服务
        finish();                                       //关闭活动
```

        }
    }

注意：

（1）隐式调用服务组件，必须在 Intent 对象里设定当前应用的包名。

（2）为简化开发，后台播放的音乐文件是应用（模块）自带的，因此，不必使用读外部存储的文件权限。将服务改写成播放外部存储的音乐文件，请读者自行完成。

5.1.5 绑定服务方式与服务代理

一个组件还可以与一个 Service 进行绑定来实现组件之间的交互，甚至可以执行 IPC（Inter-Process Communication）进程间通信。被绑定的服务依附在主进程上，而不是独立的进程，这能在一定程度上节约系统资源。

使用 bindService()方法启动 Service 的生命周期和 Activity 对象一致，也就是说，使用 bindService()方法启动 Service 后，Service 就与调用 bindService()方法的进程共存，当调用 bindService()方法的进程结束了，那么它绑定的 Service 也要跟着被结束。通常，在 Activity 的 onStop()回调函数中加上取消绑定 Service 的方法 unbindService()来停止服务连接。使用绑定服务的步骤如下：

（1）创建服务连接接口类型的对象作为类成员

绑定服务要通过服务连接接口 android.content.ServiceConnection 类型的对象获取 Service 中的状态和数据信息。接口 ServiceConnection 的定义，如图 5.1.4 所示。

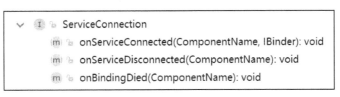

图 5.1.4　接口 ServiceConnection 的定义

注意： 接口 ServiceConnection 的前两个方法是必须重写的。

（2）在调用服务的 Activity 组件 onCreate()方法里，使用 bindService()方法启动服务。在绑定服务方法 bindService(Intent,ServiceConnection,int)中，各参数的含义如下。

第 1 个参数是一个明确指定要绑定 Service 的 Intent。

第 2 个参数是 ServiceConnection 对象。

第 3 个参数是一个标志，它表明绑定中的操作，一般应是 BIND_AUTO_CREATE，表示当 Service 不存在时创建一个，其他可选的值是 BIND_DEBUG_UNBIND 和 BIND_NOT_FOREGROUND，不想指定时设为 0 即可。

通过绑定方式使用 Service，不仅能够获取 Service 实例正常启动 Service，还能够调用 Service 中的公有方法和属性。为了使 Service 支持绑定，需要在 Service 中重载 onBind()方法，并在 onBind()方法中返回 Service 实例。这个过程就是通过 Binder（代理人）对象获取 Service 对象，进而获取所需的服务。

注意：使用 bindService()方法启动服务的方式称为绑定服务方式；在例 5.1.1 中，使用 startService()方法启动服务的方式称为非绑定服务方式。

后台播放是指在 Activity 中调用一个已经注册的 Service，这不同于前面介绍的前台播放。

【例 5.1.2】 以绑定方式调用后台音频播放服务。

（1）在项目 Example5 里，新建名为 example5_2 的模块。

（2）在模块 example5_2 的程序包名新建以绑定服务方式编写的服务程序，其代码如下：

```java
/*
    Service 是编写服务的抽象基类
    onCreate()和 onDestroy()是 Service 的 2 个重要生命周期方法，需要在其内写代码
    代理人例子："取款人—自动取款机—银行服务器"中的自动取款机
    非绑定方式启动服务时，不需要在生命周期方法 onBind()内写代码
*/
public class MyAudioService extends Service {
    private MediaPlayer mp;                        //成员
    //定义代理人内部类
    public    class PlayBinder extends Binder {    //用作代理的内部类
        public   void MyMethod(){                  //服务方法
            //应用自带的音乐文件 white.mp3
            mp=MediaPlayer.create(getApplicationContext(),R.raw.white);
            mp.start();
        }
    }
    @Override
    public IBinder onBind(Intent intent) {
        // TODO: Return the communication channel to the service.
        //throw new UnsupportedOperationException("Not yet implemented");
        return    new PlayBinder();                //返回服务代理类对象
    }
    @Override
    public void onDestroy(){                       //当服务销毁时
        if(mp!=null){
            mp.stop();                             //停止音乐播放
            mp.release();
        }
        super.onDestroy();
    }
}
```

（3）编写使用绑定服务方式的 Activity 组件，其代码如下：

```
/*
    创建接口 ServiceConnection 对象
    以绑定方式启动服务，需要建立服务代理人
    按返回值销毁活动时，也将销毁服务，从而停止音乐播放（见服务的 onDestroy()方法）
*/
public class MainActivity extends AppCompatActivity {
    @Override
    protected void onCreate(Bundle savedInstanceState) {
        super.onCreate(savedInstanceState);
        setContentView(R.layout.activity_main);
        final ServiceConnection conn=new ServiceConnection() {
            @Override
            //建立服务连接时
            public void onServiceConnected(ComponentName name, IBinder service) {
                //获取代理人对象
                MyAudioService.PlayBinder playBinder= (MyAudioService.PlayBinder) service;
                //调用代理方法
                playBinder.MyMethod();
            }
            @Override
            public void onServiceDisconnected(ComponentName name) {
                //断开服务连接
            }
        };
        Intent intent=new Intent(getApplicationContext(),MyAudioService.class); //绑定、显式调用
        /*Intent intent=new Intent("com.wust.wzx.MUSIC_PLAY_SERVICE");//需要建立意图过滤器
        intent.setPackage("com.example.example5_2");   //绑定、隐式时需要指定应用的包名*/
        //绑定服务
        bindService(intent,conn,BIND_AUTO_CREATE);
    }
    //不需要写 onDestroy()方法
}
```

5.2 远程服务

5.2.1 远程服务概念

前面介绍的是本地服务，本地服务的特点是使用的服务含于应用的内部，即不存在跨进程调用。

在 Android 系统中，每个应用程序（模块）在各自的进程中运行，这些进程之间彼此是隔离的。要完成进程之间数据（对象）的传递，需要使用 Android 支持的进程间通信（Inter-Process Communication，IPC）机制。Android 系统没有使用传统的 IPC 机制（如共享内存、管道、消息队列和 Socket 等），而是采用了 Intent 和远程服务的方式实现 IPC。

在 Android IPC 中，Binder 是进程通信的媒介，Parcel 是进程通信的内容。

注意：
（1）本地服务与远程服务不是根据距离划分的，而是根据应用的包名划分的。
（2）调用 Web 服务，本质上也是调用远程服务。

5.2.2 Android 跨进程调用与接口定义语言 AIDL

Android 系统中的不同进程之间不能直接访问相互的内存空间，为了使数据能够在不同进程间传递，数据必须先转换成能够穿越进程边界的系统级原始语言，同时，在数据完成进程边界穿越后，再转换成原有的格式。

AIDL（Android Interface Definition Language）是 Android 自定义的接口描述语言，可以简化进程间数据格式转换和数据交换的代码，通过定义 Service 内部的公有方法，允许在不同进程的调用者和 Service 之间相互传递数据。

AIDL 与 Java 的接口定义非常相似，唯一不同的是 AIDL 允许定义函数参数的传递方向，共有三种方向。标识为 in 的参数将从调用者传递到远程服务中，标识为 out 的参数将从远程服务传递到调用者中，标识为 inout 的参数先从调用者传递到远程服务中，再从远程服务返回调用者。

使用 AIDL 实现跨进程调用（远程服务）的服务器端开发，其主要步骤如下。

（1）创建接口描述文件。在 Android 项目的源文件夹的某个包里，创建扩展名为.aidl 的文件（扩展名不是.java）。在该文件中，定义了客户端可用的方法和数据的接口。如果.aidl 文件的内容是正确的，Android Studio 会在相关文件夹里自动生成一个 Java 接口文件（*.java），该接口有一个继承命名为 Stub 的内部抽象类（并且实现了一些 IPC 调用的附加方法），在 Stub 类里又含有一个名为 Proxy（代理）的内部类。

（2）建立一个服务类（Service 的子类）并实现接口。在服务类中，定义远程服务代理对象，它所在的类重写了抽象类 YourInterface.Stub 里与.aidl 文件中相对应的方法。

（3）在 Service 的绑定方法 onBind(Intent) 里返回实现了接口的实例对象。

（4）在清单文件中注册 AIDL 对应的服务。

注意： 位于系统文件夹 sdk/build-tools 内的 AIDL 编译器自动编译成接口。

使用 AIDL 跨进程调用（远程服务）的客户端开发，其主要步骤如下。

（1）创建与服务器端相同的接口描述文件。

（2）在主 Activity 中，定义远程服务连接对象，在其内创建远程服务对象。

（3）在主 Activity 中使用绑定方式调用远程服务，通过远程服务对象调用远程服务里的方法。

注意： 使用 AIDL 语言建立远程通信接口描述文件，它是服务器端与客户端共同拥有的。

5.2.3 远程服务的建立与使用实例

下面通过一个实例，说明远程服务的建立与调用。

【例 5.2.1】 以远程服务方式调用音频播放服务程序。

远程服务模块部署和运行一次后，再运行客户端模块。在客户端，单击"播放"按钮就可以远程服务调用方式来播放音乐；单击"停止"按钮，则停止远程服务。两个模块的运行界面，如图 5.2.1 所示。

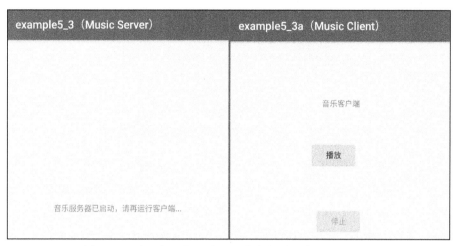

图 5.2.1　以远程服务方式调用音频播放服务的运行界面

主要设计步骤如下：

（1）在项目 Example5 里，新建名为 example5_3 的模块。

（2）在模块 example5_3 的文件夹 res 里创建文件夹 raw，粘贴一个名为 white.wp3 的音乐文件至文件夹 raw。

（3）右键服务端模块的包名，使用菜单 New→AIDL→AIDLFile→输入 IRemoteService，此时，将自动在文件夹 java 里创建文件夹 aidl，其内生成与程序包名相同的包名，在该包内包含文件 IRemoteService.aidl。注释默认创建的方法后，增加 play()和 stop()两个方法。文件 IRemoteService.aidl 的代码如下：

```
package com.example.example5_3;
interface IRemoteService {
    /*void basicTypes(int anInt, long aLong, boolean aBoolean, float aFloat,
            double aDouble, String aString);*/
    void play();    //音乐播放
    void stop();    //音乐停止
}
```

使用工具重构模块，将自动生成与文件 IRemoteService.aidl 相对应的接口文件。验证这一点，需要展开模块文件 java(generated)，如图 5.2.2 所示。

注意：在 Android 项目视图时，aidl 文件夹不可删除，切换至 Project 视图才可以。

（4）新建名为 example5_3a 的客户端模块，切换到 Project 视图，复制服务端模块的文件夹 aidl 到客户端模块的相应位置后，使用工具重构本模块，再切换项目视图为 Android 视图，可以看到生成了与服务端模块相同的接口及其包名。

图 5.2.2　服务端项目 AIDL 文件及其对应的接口文件

（5）在服务端模块的程序包里，创建名为 MusicPlayService 的服务组件，定义一个 MediaPlayer 类型的对象 mPlayer 和一个接口内部类 IRemoteService.Stub 的对象 iBinder 作为类成员。其中，IRemoteService.Stub 为抽象类，因此，定义作为代理人的 iBinder 对象时，需要重写与接口方法同名的两个方法，分别对 mPlayer 应用方法 start()和 stop()。

在 MusicPlayService 的 onCreate()方法里，实例化对象 mPlayer；在 onBind()方法里，返回代理人对象 iBinder；在 onUnbind()方法里，应用方法 mPlayer.stop()。

服务端模块的服务组件 MusicPlayService 的代码如下：

```
/*
    创建代理人：为屏蔽客户调用远程主机上的对象，必须提供某种方式来模拟本地对象，
    这种本地对象称为存根（stub），存根负责接收本地方法调用，
         并将它们委派给各自的具体实现对象
*/
public class MusicPlayService extends Service {
    MediaPlayer mPlayer;
    //创建代理人对象 iBinder
    IBinder iBinder = new IRemoteService.Stub() {
        //接口的内部类是一个抽象类
        //IRemoteService.Stub 又包含一个名为 Proxy 的代理类，
        //定义了两个与接口同名的方法 play()和 stop()
        @Override
        public void play() throws RemoteException {
            if(!mPlayer.isPlaying()){
                mPlayer.start();
                Log.d("测试", "play ");
            }
        }
        @Override
        public void stop() throws RemoteException {
            if(mPlayer.isPlaying()){
                mPlayer.stop();
                Log.d("测试", "stop ");
```

```
                try{
                    mPlayer.prepare();
                }catch (IOException e){
                    e.printStackTrace();
                }
                mPlayer.seekTo(0);
            }
        }
    };
    @Override
    public void onCreate() {
        super.onCreate();
        if(mPlayer==null){
            mPlayer =MediaPlayer.create(this, R.raw.white);     //创建对象
            mPlayer.setLooping(true);
            Log.d("测试", "created");
        }
    }
    @Override
    public IBinder onBind(Intent intent) {                      //服务绑定时
        Toast.makeText(this, "服务已绑定！", Toast.LENGTH_SHORT).show();
        return iBinder;                                         //返回代理人
    }
    @Override
    public boolean onUnbind(Intent intent) {
        if (mPlayer!=null){
            mPlayer.stop();
            mPlayer.release();
            Log.d("测试", "onUnbind");
        }
        return super.onUnbind(intent);
    }
}
```

（6）在清单文件自动生成的服务组件注册的标签里，增加意图过滤器的代码（加粗文本）如下：

```xml
<service android:name=".MusicPlayService"
    android:process=":remote">
    <intent-filter>
        <action android:name="com.example.services.MusicService"/>
    </intent-filter>
</service>
```

（7）在客户端模块的布局文件里，添加两个分别表示播放和停止的 Button 控件。

（8）在客户端模块的 MainActivity 里，定义一个接口 IRemoteService 类型的对象 mServer 和一个接口 ServiceConnection 类型的对象 conn 作为类成员。其中，定义 conn

时，在方法 onServiceConnected()里，使用 IRemoteService.Stub.asInterface()实例化对象 mServer。

在 onCreate()里，创建一个以隐式调用服务组件的意图对象 intent，使用服务绑定方法 bindService(intent,conn,BIND_AUTO_CREATE)；在播放按钮的单击事件监听代码里，使用方法 mServer.play()；在停止按钮的单击事件监听代码里，使用方法 mServer.stop()。

客户端模块 Activity 组件程序 MainActivity 的代码如下：

```java
/*
    音乐客户端，以远程方式调用音乐服务端
    需要使用"绑定+隐式"远程服务调用
    IRemoteService 是服务端及客户端的公共接口，由相同的 AIDL 文件（含包名）自动生成
 */
public class MainActivity extends AppCompatActivity implements View.OnClickListener{
    Button btn1,btn2;
    //服务端项目包名，需要查看服务端项目清单文件
    static final String PACKAGE="com.example.example5_3";
    //服务端项目里定义的意图动作名，需要查看服务端项目清单文件
    static final String ACTION = "com.example.services.MusicService";
    boolean isBinded=false;
    IRemoteService mServer;                    //关键
    ServiceConnection conn = new ServiceConnection() {
        @Override
        public void onServiceConnected(ComponentName name, IBinder service) {
            //建立服务连接时的远程服务代理对象
            mServer = IRemoteService.Stub.asInterface(service);
            isBinded=true;
        }
        @Override
        public void onServiceDisconnected(ComponentName name) {
            isBinded=false;
            mServer=null;
        }
    };
    @Override
    protected void onCreate(Bundle savedInstanceState) {
        super.onCreate(savedInstanceState);
        setContentView(R.layout.activity_main);
        btn1=findViewById(R.id.btn1);btn1.setOnClickListener(this);
        btn2=findViewById(R.id.btn2);btn2.setOnClickListener(this);
        Intent intent = new Intent(ACTION);
        intent.setPackage(PACKAGE);      //隐式调用需要指定服务端项目的包名
        bindService(intent,conn,BIND_AUTO_CREATE);
    }
    @Override
    public void onClick(View v) {            //监听器接口方法
```

```
            int id = v.getId();
            switch (id){
                case R.id.btn1:
                    try {
                        mServer.play();        //调用服务接口方法
                        btn1.setEnabled(false);btn2.setEnabled(true);
                    } catch (RemoteException e) {
                        e.printStackTrace();
                    }
                    break;
                case R.id.btn2:
                    try {
                        mServer.stop();        //调用服务接口方法
                        btn2.setEnabled(false);btn1.setEnabled(true);
                    } catch (RemoteException e) {
                        e.printStackTrace();
                    }
            }
        }
        @Override
        protected void onDestroy() {
            if(isBinded){
                unbindService(conn);           //取消绑定
                mServer = null;
                isBinded=false;
            }
            super.onDestroy();
        }
    }
```

（9）先部署服务端模块，再部署客户端模块，然后做运行测试。

注意：音乐服务器仅需要部署、运行一次。

5.3 综合应用实例——自动挂断来电后回复短信

在 Android 系统中，以程序方式挂断或应答打进来的电话，需要使用 Android 系统提供的电话服务接口 com.android.internal.ITelephony。使用 import 指令导入这个包，在通常情况下会失败。

Android 系统存在很多像 ITelephony 这样隐藏的接口，即未开放的 Android 接口，因此，不能在 SDK 中找到。事实上，接口 ITelephony 的包名 com.android.internal 就表明了这个意思。由于 android.telephony.TelephonyManager 提供了返回类型为 ITelephony 的私有方法 getITelephony()，因此，使用 Java 的反射机制，就可以间接获取接口 ITelephony 类型的对象。为了实现跨进程的方法调用（ITelephony 提供了 endCall()等方法），需要

AIDL 来配合。

【例 5.3.1】 自动挂断来电后回复短信。

设计思想：在自定义类 Constant 定义两个静态属性，分别保存防打扰开关与自动回复短信的内容。在 MainActivity 中，获取服务所需的数据，包括是否开启免打扰和回复短信的文本，并保存于类 Constant 的两个静态属性里。在服务程序中，使用保存在类 Constant 里的属性值，通过 AIDL 协议与 Java 反射获得 iTelephony 对象，调用其 endCall() 方法挂断电话；使用 SmsManager 对象的 sendTextMessage()方法发送短信。

程序运行：将本应用项目部署到一部手机上，单击文本"开启防打扰模式"项，可以修改回复短信的内容，效果如图 5.3.1 所示。

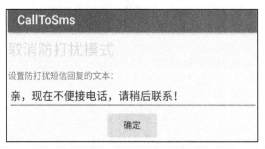

图 5.3.1　程序运行效果

在另一部手机上，向第一部手机打电话，则电话自动被挂断，稍后收到来自第一部手机的短信通知。同时，在第一部手机的通知栏里也会出现未接电话的通知。

主要设计步骤如下。

（1）在项目 Example5 里，创建一个名为 example5_4 的模块。

（2）右击模块名→New→Folder→AIDL Folder，创建名为 aidl 的空文件夹。右击文件夹 aidl→New→package，输入特定的包名：com.android.internal.telephony。右击包名→New→File，输入特定的文件名：ITelephony.aidl，然后复制如下代码：

```
package com.android.internal.telephony;
interface ITelephony{
    boolean endCall();
    void answerRingingCall();
}
```

注意：创建 AIDL 文件更加快捷的操作方法是，先在程序文件夹 java 里创建一个包名 com.android.internal.telephony，然后右击该包名，选择 New→AIDL→AIDL File→输入 ITelephony。最后，删除 java 文件夹里的包名 com.android.internal.telephony。

使用工具 ⚒ 重构模块，以便自动生成 AIDL 文件对应的接口文件。

查验生成是否成功，需要切换项目视图，如图 5.3.2 所示。

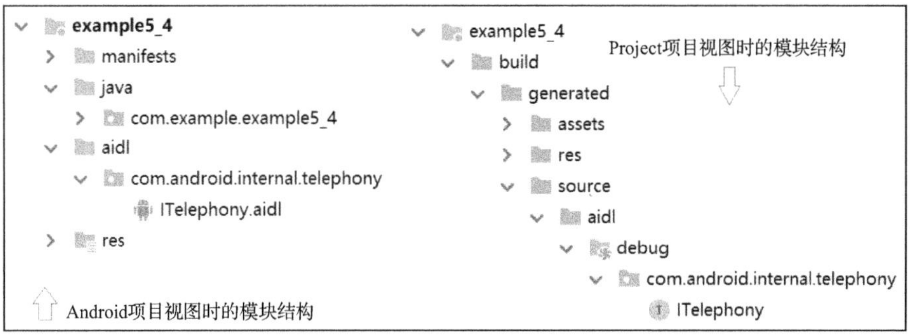

第 5 章 服务组件及其应用

```
v example5_4                                  v example5_4
  > manifests                                   v build
  v java                                           v generated         Project项目视图时的模块结构
      > com.example.example5_4                      > assets
  v aidl                                            > res
      v com.android.internal.telephony            v source
          ITelephony.aidl                          v aidl
  > res                                              v debug
                                                     v com.android.internal.telephony
          Android项目视图时的模块结构                          ITelephony
```

图 5.3.2　含有 AIDL 文件及其对应接口文件的项目文件系统

（3）在模块的程序包里，创建名为 PhoneService 的服务组件，其代码如下：

```
/*
    本服务程序包含了电话状态监听：来电时自动挂断电话
    自动挂断电话后，发送短信，其内容来源于一个文本文件
    需要动态申请的危险权限组：电话、外短信和外部存储
    关键知识点：定义 AIDL 文件生成相应的接口文件、使用 Java 反射机制创建 ITelephony 对象
*/
public class PhoneService extends Service {
    private ITelephony iTelephony;          //通信接口对象，需要先建立相应的 AIDL 文件
    private TelephonyManager manager;       //通信管理器
    private String inNumber = null;         //记录来电号码，供发短信时使用
    public void onCreate() {
        if(Constant.flag) {
            getphoner();                    //通过反射获得 iTelephony 对象
            manager.listen(new PhoneStateListener() {    //PhoneStateListener 是类而不是接口
                @Override
                public void onCallStateChanged(int state, String incomingNumber) {
                    super.onCallStateChanged(state, incomingNumber);
                    inNumber = incomingNumber;       //来电号码保存，供发通知时使用
                    switch (state) {
                        case 1:   //当来电时
                            try {
                                if(Constant.flag) {    //当开启免打扰时
                                    //通过 iTelephony 对象，调用接口 ITelephony 的方法
                                    iTelephony.endCall();
                                    Log.i("wzxtest",incomingNumber);
                                    Toast.makeText(getApplicationContext(), "endcall:" +
                                            incomingNumber, Toast.LENGTH_SHORT).show();
                                    sendSMS(incomingNumber, "<CallToSms 自动回复>\n" +
                                            Constant.anti_sms);        //发送短信
                                    //可增加代码在通知栏提示短信发送成功
                                }
                            } catch (RemoteException e) {
                                e.printStackTrace();
```

```java
                    }
                    break;
                default:
                    break;
            }
        }
    }, PhoneStateListener.LISTEN_CALL_STATE);
}
//反射得到接口 ITelephony 类型的对象 iTelephony
public void getphoner(){
    //在 Service 组件里调用系统服务,因为 Service 是 Context 的子类
    manager = (TelephonyManager)getSystemService(TELEPHONY_SERVICE);
    //创建与 TelephonyManager 关联的 Class 类型对象
    Class<TelephonyManager> c = TelephonyManager.class;
    Method method = null;
    try {
        //类 TelephonyManager 的方法 getITelephony()
        method = c.getDeclaredMethod("getITelephony", (Class[])null);
        //设置允许访问类的私有方法
        method.setAccessible(true);
        //(方法)反射得到该方法返回类型的对象
        // 需要先建立 AIDL 文件生成接口 ITelephony,因为 ITelephony 是 Android 未公开的接口
        iTelephony = (ITelephony) method.invoke(manager, (Object[])null);
    } catch (IllegalArgumentException e) {
        e.printStackTrace();
    } catch (Exception e) {
        e.printStackTrace();
    }
}
public void sendSMS(String phonenumber, String msg){ //下面的发送短信方法,需要短信权限
    SmsManager smsManager = SmsManager.getDefault();
    // 将短信内容分段,装入 ArrayList
    ArrayList<String> list = smsManager.divideMessage(msg);
    for (String sms : list)                              //分段发送
        smsManager.sendTextMessage(phonenumber, null, sms, null, null);
}

public int onStartCommand(Intent intent, int flags, int startId) {
    return super.onStartCommand(intent, flags, startId);
}
@Override
public IBinder onBind(Intent arg0) {
    return null;
}
}
```

注意：编写本服务组件的难点是对 Android 未公开接口 ITelephony 的使用。在 AIDL 文件生成接口后，使用 Java 反射机制，对 TelephonyManager 的方法 getITelephony()进行反射调用，创建 ITelephony 的实例对象。

（4）在模块的程序包里，创建 Activity 组件程序 MainActivity，其代码如下：

```
/*
    需要动态申请的危险权限组有 3 个：电话、短信和外部存储
    打开免打扰模式时，需要设置短信文本
    当设置免打扰模式时，才可编辑用于发短信的文本
*/
public class MainActivity extends AppCompatActivity {
    private TextView textView1;                         //用于设防或取消
    private TextView textView2;                         //提示文本
    private String fileName = "anti_disturb_sms.txt";   //短信文本文件，可修改
    private String message;                             //短信内容
    private EditText editText;                          //短信内容编辑文本框
    private Button button;
    Intent intent;
    //需要多个危险权限数组的动态申请
    private static String[] permissions = {"android.permission.READ_PHONE_STATE",
                                "android.permission.SEND_SMS",
                                "android.permission.WRITE_ EXTERNAL_ STORAGE"};
    @Override
    protected void onCreate(Bundle savedInstanceState) {
        super.onCreate(savedInstanceState);
        setContentView(R.layout.activity_main);
        List<String> mPermissionList = new ArrayList<>(); //存放未授权的权限
        mPermissionList.clear();
        for (int i = 0; i < permissions.length; i++) {
            if (ContextCompat.checkSelfPermission(this, permissions[i]) !=
                                        PackageManager.PERMISSION_GRANTED) {
                mPermissionList.add(permissions[i]);
            }
        }
        //Build.VERSION.SDK_INT 表示 Android 设备的 API 版本，API 23 对应于 Android 6.0
        if (Build.VERSION.SDK_INT < Build.VERSION_CODES.M) {   //Android 6.0 以下版本时
            fun();                                  //做该做的
        } else {
            if (mPermissionList.isEmpty()) {
                fun();                              //应用已经获得权限时直接执行
            } else{
                //集中请求未授权的权限；将 List 转为数组
                String[] needPermissions = mPermissionList.toArray(new String[mPermissionList.size()]);
                //请求权限后执行回调方法
                ActivityCompat.requestPermissions(this, needPermissions, 1);
```

```java
            }
        }
    }
    public void fun(){
        textView1 = findViewById(R.id.textView1);
        textView2 = findViewById(R.id.textView2);
        editText = findViewById(R.id.editText);
        button = findViewById(R.id.button);
        textView1.setOnClickListener(new View.OnClickListener() {   //监听：设置或取消防打扰
            public void onClick(View view) {
                if(Constant.flag) {
                    Constant.flag = false;                          //取消防打扰
                    textView1.setText(R.string.ennable_anti_disturb);
                    textView1.setTextColor(Color.parseColor("#000000"));//关闭状态字体为黑色
                    textView2.setVisibility(View.INVISIBLE);
                    editText.setVisibility(View.INVISIBLE);button.setVisibility
                                                            (View.INVISIBLE); //隐藏
                } else {
                    Constant.flag = true;                           //设置防打扰
                    textView1.setText(R.string.cancel_anti_disturb);
                    //开启状态时将文字颜色改为粉红色
                    textView1.setTextColor(Color.parseColor("#ffcccc"));
                    textView2.setVisibility(View.VISIBLE);
                    editText.setVisibility(View.VISIBLE);button.setVisibility
                                                            (View.VISIBLE); //可见
                    try{
                        message = buildSms(fileName); //创建短信内容
                        Constant.anti_sms=message;     //也是服务程序里发送的短信内容
                        editText.setText(message);
                        button.setOnClickListener(new View.OnClickListener(){
                            @Override
                            public void onClick(View arg0) {
                                if("".equals(editText.getText().toString())) {
                                    Toast.makeText(getApplicationContext(),
                                            "回复为空！请重设！",
                                                        Toast.LENGTH_SHORT).show();
                                } else {
                                    message = editText.getText().toString();
                                    try {
                                        writeFile(fileName, message);
                                        Constant.anti_sms=message;
                                    }
                                    catch (IOException e) {
                                        e.printStackTrace();
                                    }
                                }
```

```java
                    }
                });
                // 设置防打扰模式后，以"隐式+非绑定"方式服务调用
                intent = new Intent("com.wzx.service.ACTION_MYSERVICE");
                //隐式配置
                intent.setPackage("com.example.example5_4");
                startService(intent);
            }
            catch(Exception e){
                e.printStackTrace();
            }
        }
    }
});
}
@Override
public void onRequestPermissionsResult(int requestCode, @NonNull String[] permissions,
                                        @NonNull int[] grantResults) {
    super.onRequestPermissionsResult(requestCode, permissions, grantResults);
    switch (requestCode) {
        case 1:
            for (int i = 0; i < grantResults.length; i++) {
                if (grantResults[i] != PackageManager.PERMISSION_GRANTED) {
                    Toast.makeText(this, "未授权，无法实现全部功能！",
                            Toast.LENGTH_SHORT).show();
                    finish();
                }
            }
            fun();                                  //全部授权时执行
    }
}
public String buildSms(String fileName) throws IOException {    //读取文件 anti_disturb_ sms.txt
    String res = "";
    try{
        String basePath = Environment.getExternalStorageDirectory().getPath();
        File file = new File(basePath+"/"+fileName);
        if(!file.exists()) {                            //回复短信文件不存在时
            file.createNewFile();
            //首次安装使用时默认回复的短信内容，修改后写入 SD 卡
            res = "亲，现在不便接电话，请稍后联系！";
            writeFile(fileName,res);                //调用
        } else {
            FileInputStream fis = new FileInputStream(file);
            int length = fis.available();
            byte [] buffer = new byte[length];
            fis.read(buffer);
```

```
                    //res = EncodingUtils.getString(buffer, "UTF-8");        //API 21 之后不用
                    res=new String(buffer,"UTF-8");    //
                    fis.close();
                }
        }catch(Exception e){
                e.printStackTrace();
        }
        return res;
    }
    public void writeFile(String fileName ,String write_str) throws IOException { //创建短信文件
            String basePath = Environment.getExternalStorageDirectory().getPath();
            File file = new File(basePath+"/"+fileName);
            Log.i("wzxtest",basePath+"/"+fileName);
            FileOutputStream fos = new FileOutputStream(file);
            byte[] bytes = write_str.getBytes();
            fos.write(bytes);
            fos.close();
            Toast.makeText(this, "写入成功", Toast.LENGTH_SHORT).show();
    }
}
```

习 题 5

一、判断题

1. Android 系统允许其他组件通过 Intent 对象启动 Service 组件。
2. 隐式调用服务组件，必须在 Intent 对象里设定当前应用的包名。
3. Service 组件的生命周期方法与调用方式无关。
4. 本地服务调用，并不需要 AIDL。
5. Activity 组件调用远程服务组件，只能以绑定方式进行。
6. Android 远程服务中的服务组件，可以被客户端显式调用。
7. 在项目通常的 Android 方式下，可以删除 aidl 文件夹。

二、选择题

1. 非绑定服务的主要业务逻辑代码应出现在服务组件的_____方法里。
   A．onUnbind()　　B．onCreate()　　C．onDestroy()　　D．onBind()
2. 使用绑定服务时，在 Activity 组件里必须重写_____方法。
   A．startActivity()　　　　　　　B．onServiceConnected()
   C．onActivityResult()　　　　　D．onPause()
3. 被绑定的服务，需要在服务的_____方法里返回一个 Binder 对象。
   A．onBind()　　B．onStart ()　　C．onUnbind()　　D．onDestroy()
4. 发送短信时，其内容是由_____对象传递的。
   A．Service　　　　　　　　　　B．BroadcastReceiver
   C．Intent　　　　　　　　　　　D．ContentProvider
5. 在 Android 程序设计时，不可直接导入使用的接口是_____。
   A．SharedPreferences　　　　　B．IBinder
   C．ServiceConnection　　　　　D．ITelephony ()

三、填空题

1. Service 组件与 Activity 组件一样，含于相同的软件包_____里。
2. 服务有显式启动和隐式启动两种。若服务和调用服务的组件在不同的应用程序中，则应使用_____启动。
3. 定义被绑定的服务时，其 onBind()方法返回的接口类型是_____。
4. 建立远程服务调用，需要使用_____语言来描述和生成远程服务接口。
5. 通过方法 bindService()绑定 Service 时，Service 组件最常用的生命周期方法 onCreate()和_____将被调用。

实 验 5

一、实验目的

（1）理解 Service、BroadcastReceiver 与 Activity 的不同点。
（2）掌握 Service 两种调用方式（显式与隐式）的用法。
（3）掌握 Service 两种启动方式（绑定与非绑定）的用法。
（4）掌握本地服务调用与远程服务调用的用法。

二、实验内容及步骤

【预备】在 Android Studio 中，新建名为 Example5 的项目后，访问本课程配套的网站 http://www.wustwzx.com/as/sy/sy05.html，复制相关代码，完成如下几个模块的设计。

1．调用后台音频播放服务（参见例 5.1.1）

（1）在项目 Example 5 里，新建名为 example5_1 的模块。
（2）复制一个音乐文件至文件夹 res/raw 里，并命名为 white.mp3。
（3）在程序文件夹里，使用菜单 File→New→Service，创建名为 MyAudioService 的服务组件，定义 MediaPlayer 类型的对象 mp 作为类成员。
（4）在 MyAudioService 的 onCreate()方法里，创建对象 mp 并应用 start()方法。
（5）在 MyAudioService 的 onDestroy()方法里，使用 mp.stop()方法。
（6）在布局文件里添加两个表示播放和停止的 Button 按钮。
（7）在类 MainActivity 里，分别定义 1 个 Intent 类型的对象 intent 和两个 Button 类型的对象作为类成员；在 onCreate()方法里，实例化两个按钮和 1 个显式启动 MyAudioService 的对象 intent；监听播放按钮并在 onClick()方法里使用方法 startService(intent)；监听停止播放按钮并在 onClick()方法里使用方法 stopService(intent)。
（8）在类 MainActivity 的 onDestroy()方法里，依次使用应用方法 stopService(intent) 和 finish()。
（9）先部署本模块，然后，在清单文件里定义服务组件的意图过滤器，并在 MainActivity 里使用隐式调用，再做运行测试。

2．以绑定服务方式调用音频播放服务（参见例 5.1.2）

（1）在项目 Example 5 里，新建名为 example5_2 的模块。
（2）复制一个音乐文件至文件夹 res/raw 里，并命名为 white.mp3。
（3）在程序文件夹里，使用菜单 File→New→Service，创建名为 MyAudioService 的服务组件，定义 MediaPlayer 类型的对象 mp 作为类成员。
（4）在类 MyAudioService 里，定义一个继承 Binder 的子类 PlayBinder 作为内部类。在类 PlayBinder 里定义一个播放音乐的方法 MyMethod()。

（5）在类 MyAudioService 的 onBind ()方法里，返回一个 PlayBinder 类型的对象。

（6）在类 MyAudioService 的 onDestroy()方法里，编写停止音频播放的代码。

（7）在类 MainActivity 里，定义一个接口 ServiceConnection 类型的对象 conn，在其方法 onServiceConnected()里，获得 PlayBinder 类型的对象，并使用其方法 MyMethod()。

（8）在类 MainActivity 的 onCreate()方法里，创建一个使用服务组件的意图对象 intent，使用服务绑定方法 bindService(intent,conn,BIND_AUTO_CREATE)。

（9）部署本模块做运行测试。

3．以远程服务方式调用音频播放服务程序（参见例 5.2.1）

（1）在项目 Example 5 里，新建拟作为服务端、名为 example5_3 的模块，复制一个音乐文件至文件夹 res/raw 里，并命名为 white.mp3 后，使用工具 🔄 做模块资源同步操作。

（2）右击模块名，使用菜单 New→AIDL 创建一个名为 IRemoteService 的 AIDL 文件，定义两个方法 void play()和 void stop()，使用工具 🔨 重构模块。

（3）在项目 Example 5 里，新建拟作为客户端、名为 example5_3a 的模块。切换项目视图为 Project 视图，复制 example5_3 模块的文件夹 src/main/aidl 到 example5_3a 模块的相应位置后，使用工具 🔨 重构本模块，再切换项目视图为 Android 视图。

（4）在类 MyAudioService 里，定义一个 MediaPlayer 类型的对象 mPlayer 和一个接口内部类 IRemoteService.Stub 的对象 iBinder 作为类成员。其中，IRemoteService.Stub 为抽象类，因此，定义作为代理人的 iBinder 对象时，需要重写与接口方法同名的两个方法，分别对 mPlayer 应用方法 start()和 stop()。

（5）在类 MyAudioService 的 onCreate()方法里，实例化对象 mPlayer；在 onBind()方法里，返回代理人对象 iBinder；在 onUnbind()方法里，应用方法 mPlayer.stop()。

（6）在客户端模块的布局文件里，添加两个表示播放和停止的 Button 控件。

（7）在客户端模块的 MainActivity 里，定义一个接口 IRemoteService 类型的对象 mServer 和一个 ServiceConnection 类型的对象 conn 作为类成员。其中，定义 conn 时，在方法 onServiceConnected()里，使用 IRemoteService.Stub.asInterface()实例化对象 mServer。

（8）在客户端模块的 MainActivity 的 onCreate()里，创建一个以隐式调用服务组件的意图对象 intent，使用服务绑定方法 bindService(intent,conn,BIND_AUTO_CREATE)；在播放按钮的单击事件监听代码里，使用方法 mServer.play()；在停止按钮的单击事件监听代码里，使用方法 mServer.stop()。

（9）先部署服务端模块，然后部署客户端模块，再做运行测试。

4．组件综合使用——自动挂断来电后回复短信程序设计（参见例 5.3.1）

（1）在项目 Example 5 里，新建名为 example5_4 的模块，在清单文件里配置本应用所需要电话组的 2 个相关权限：发短信权限和读/写外部存储权限。

（2）右击模块名，创建 aidl 文件夹，在 aidl 里创建名为 com.android.internal.telephony

的包，在包里创建名为 ITelephony.aidl 的文件，使用工具重构模块。

（3）在模块程序包内，创建名为 Constant 的类，包含 2 个静态成员常量用于实现 Activity 组件与 Service 组件之间的信息共享。

（4）在模块程序包内，创建名为 PhoneService 的 Service 组件，在清单文件注册该组件的代码内，使用标签<intent-filter>并内嵌标签<action>来添加意图过滤器代码，指定隐式调用的服务名称 com.example.services.MusicService，它将在 Activity 组件里使用。

（5）在组件程序 PhoneService 里，依次定义接口 ITelephony 类型的对象 iTelephony、TelephonyManager 类型的对象 manager 和 String 类型的对象 inNumber 作为类成员。

（6）在组件程序 PhoneService 的 onCreate()方法里，实例化对象 manager；使用 Java 反射技术获取，实例化对象 iTelephony；使用方法 manager.listen()在电话到来时，实例化对象 inNumber；使用方法 iTelephony.endCall()挂断电话、自动回复短信。

（7）在组件程序 MainActivity 里，依次定义 TextView 控件对象 textView1（提示和切换当前是否开启防打扰状态）、EditText 控件对象（用于）editText 和 Button 控件对象 button（执行短信文本写入）等作为类成员。

（8）在组件程序 MainActivity 的 onCreate()里，依次进行动态权限申请、监听 textView1 的状态。当处于防打扰模式时，实现短信文本的编辑及保存工作，以"隐式+非绑定"方式启动自定义的 Service 组件。

（9）部署本模块并做运行测试。

三、实验小结及思考

（由学生填写，重点写上机中遇到的问题。）

第 6 章 广播组件与通知

广播接收组件 BroadcastReceiver 也是 Android 的四大组件之一，是对广播进行过滤并响应的程序。与 Activity 和 Service 组件一样，BroadcastReceiver 组件在使用前通常也需要在清单文件里注册（BroadcastReceiver 还可以在程序里动态注册），并由 Intent 对象去激活。本章学习要点如下：
- 了解 Android 的广播机制；
- 了解 Android 广播与通知的关系；
- 掌握广播接收者的两种注册方式；
- 掌握 BroadcastReceiver 程序的编写方法。

6.1 广播与 BroadcastReceiver 组件

6.1.1 Android 广播机制

在 Android 系统中，广播（Broadcast）是一种广泛运用在应用程序之间传输信息的机制。

根据广播的来源，Android 广播可分为系统广播与用户自定义广播两种。来电话、来短信、手机没电等系统发送的消息，都属于系统广播。当某种特定的事件发生时，Android 系统都会产生一个 Intent 对象自动进行广播。

Android 系统中的广播，都由特定的广播接收者来接收、处理。广播发出后，广播接收者会根据广播的类型进行匹配，来判断是否接收该广播。例如，使用类常量 android.provider.Telephony.SMS_RECEIVED 来表示手机有短信到来时所产生的广播。只要广播接收者在清单文件里注册接收广播的类型与手机短信广播的类型相同，在手机短信到来时，该广播接收者将会获取短信广播。

一个应用程序自定义的消息，并将这些消息通过广播方式通知给特定的应用程序，属于自定义广播。

注意：

（1）Android 系统产生的广播，只有拥有相应权限的广播接收者程序才能接收到。

（2）广播都是通过 Intent 对象发送的，因此，从广播意图对象里，可获取广播携带的数据。

6.1.2 使用 BroadcastReceiver 组件定义广播接收者

广播接收组件 BroadcastReceiver 主要提供了处理广播的抽象方法 onReceive，如图 6.1.1 所示。

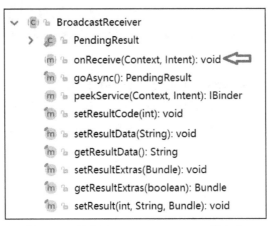

图 6.1.1 抽象类 BroadcastReceiver 的定义

广播接收者作为抽象类 BroadcastReceiver 的子类，需要重写 onReceive()方法。一个 BroadcastReceiver 对象的生命周期从调用 onReceive(Context,Intent)方法开始，到该方法返回结束。

在 Android Studio 中，使用菜单 File→New→Other→BroadcastReceiver，出现创建广播接收者类的向导。创建完成后，自动在项目清单文件里注册该组件。开发者还需要在注册广播接收者的代码添加一个意图过滤器，指定要处理的广播，示例代码如下：

```
<receiver
    android:name=".MyReceiver"
    android:enabled="true"
    android:exported="true">
    <intent-filter>
        <action android:name="android.provider.Telephony.SMS_RECEIVED" />
    </intent-filter>
</receiver>
```

注意：

（1）如果不为广播接收者程序配置过滤器，将不会接收到任何广播。

（2）在<intent-filter>标签里，可以使用多个<action>标签定义需要处理的多种广播。

Android 系统产生广播后，就会创建相应的 BroadcastReceiver 对象，在执行 onReceive()方法后，销毁该实例对象。例如，手机 Android 系统接收短信广播后，内置的广播接收者程序执行 onReceive()的结果是产生短信铃声并发出通知。

Android API 提供的 android.content.BroadcastReceiver 是广播接收者类的抽象父类。

第6章 广播组件与通知

在 Android 应用中,每个广播接收者类都使用标签<receiver>在项目的清单文件中注册,只有具备相应权限的广播接收者才能接收广播、获取 Intent 对象中的数据。

下面,通过实例说明广播接收者的使用方法。

【例6.1.1】 手机短信收到时系统广播启动广播接收者程序。

(1)复制音乐文件 white.mp3 到文件夹 res/raw 里。

(2)在清单文件中注册接收系统短信广播的权限,代码如下:

```xml
<uses-permission android:name="android.permission.RECEIVE_SMS"/>
```

新建继承 android.content.BroadcastReceiver 的类文件 SmsReceiver.java,其代码如下:

```java
public class SmsReceiver extends BroadcastReceiver {
    public SmsReceiver() { }
    @Override
    public void onReceive(Context context, Intent intent) {
        MediaPlayer.create(context,R.raw. white).start();
    }
}
```

(3)在清单文件中注册接收系统短信广播的广播接收者,代码如下:

```xml
<receiver
    android:name="com.example.simplebroadcastreceiver.SmsReceiver"
    android:enabled="true"
    android:exported="true" >
    <intent-filter>
        <action android:name="android.provider.Telephony.SMS_RECEIVED"/>
    </intent-filter>
</receiver>
```

(4)MainActivity 包含了动态申请短信接收权限和按返回键时取消广播注册等功能,其代码如下:

```java
/*
    以静态注册方式(在清单文件注册),使用系统短信广播示例
    功能:接收短信广播后,自行播放音乐
    测试时注意:短信从发送到接收,存在延时;需要运行本活动
*/
public class MainActivity extends AppCompatActivity {
    @Override
    protected void onCreate(Bundle savedInstanceState) {
        super.onCreate(savedInstanceState);
        setContentView(R.layout.activity_main);
        if (ContextCompat.checkSelfPermission(this, Manifest.permission.RECEIVE_SMS) !=
                                    PackageManager.PERMISSION_GRANTED){
            ActivityCompat.requestPermissions(this,new String[]{"android.permission.
                                                        RECEIVE_SMS"},1);
```

```
            //小米 5s（7.0 版本）不会立即弹出权限申请窗口，而是在收到短信时弹窗申请权限
            //因此，本模块不用写回调方法 onRequestPermissionsResult()
        }
    }
        ComponentName receiver = new ComponentName(this, SMSReceiver.class);
            PackageManager pm = getPackageManager();
            // 启用清单文件中定义的 SmsReceiver
            pm.setComponentEnabledSetting(receiver, PackageManager.COMPONENT_ENABLED_STATE_ENABLED, PackageManager.DONT_KILL_APP);
        @Override
        protected void onDestroy() {                                    //按手机返回键时触发
            super.onDestroy();
            //创建组件对象
            ComponentName receiver = new ComponentName(this,SMSReceiver.class);
            //获取包管理器对象
            PackageManager pm = getPackageManager();
            //禁用一个静态注册的广播接收者
            pm.setComponentEnabledSetting(receiver, PackageManager.COMPONENT_ENABLED_STATE_DISABLED, PackageManager.DONT_KILL_APP);
        }
}
```

【例 6.1.2】 收到手机短信时系统广播启动音乐播放服务程序。

【程序运行】本 Android 项目部署到手机后，此时"Stop Music"按钮不可用，按返回键可退出。当手机有短信到来时，所产生的系统广播会同时激活播放后台音乐的服务程序和应用的主 Activity，此时，可单击"Stop Music"按钮，停止音乐播放的后台程序并回到手机桌面。

与例 6.1.1 相比，只是广播接收者程序的 onReceive()方法里，同时启动了播放音乐的服务程序和 MainActivity 程序。

播放音乐服务程序的代码如下：

```
public class MyAudioService extends Service {
    MediaPlayer mediaPlayer;
    @Override
    public IBinder onBind(Intent intent) {
        return null;
    }
    @Override
    public void onCreate() {
        mediaPlayer = MediaPlayer.create(this, R.raw.black);
        mediaPlayer.start();
    }
    @Override
    public void onDestroy() {
        mediaPlayer.stop();
```

    }
}

活动程序的代码如下：

```java
/*
    当收到短信时，启动音乐播放服务，激活一个 Activity 并提供停止按钮
    测试本模块，需要卸载 example6_1 模块，不然有干扰（两个应用都接收到了短信并处理）
*/
public class MainActivity extends AppCompatActivity {
    private Button btnStop;
    private boolean isCast;                    //是否为广播激活
    @Override
    protected void onCreate(Bundle savedInstanceState) {
        super.onCreate(savedInstanceState);
        setContentView(R.layout.activity_main);
        if (ContextCompat.checkSelfPermission(this, Manifest.permission.RECEIVE_SMS) !=
                                            PackageManager.PERMISSION_GRANTED){
            ActivityCompat.requestPermissions(this,new String[]{"android.permission.
                                            RECEIVE_SMS"},1);
        }
        btnStop=findViewById(R.id.btnStop);
        Intent intent = getIntent();           //获取广播意图对象
        isCast = intent.getBooleanExtra("iscast", false);
        btnStop.setEnabled(isCast);            //设置停止按钮可用和单击监听
        btnStop.setOnClickListener(new View.OnClickListener() {
            @Override
            public void onClick(View v){
                //显式服务调用意图（非绑定式）
                Intent intent=new Intent(MainActivity.this,MyAudioService.class);
                //在 Activity 组件里，停止音乐播放服务
                MainActivity.this.stopService(intent);
                finish();                      //销毁本活动
            }
        });
    }
}
```

广播接收者组件 SMSReceiver 分别调用了模块的 MainActivity 组件和播放音乐的服务组件 MyAudioService，其代码如下：

```java
public class SMSReceiver extends BroadcastReceiver {
    @Override
    public void onReceive(Context context, Intent intent) {
        Intent serviceIntent = new Intent(context, MyAudioService.class);
        //在广播组件里，通过上下文对象启动音乐播放服务组件
        context.startService(serviceIntent);
        //新建调用 Activity 组件的意图
```

```
        Intent activityIntent = new Intent(context, MainActivity.class);
        activityIntent.putExtra("iscast", true);           //携带数据
        //新建栈用来存放被启动的 Activity（当已经存在时，只做移动处理）
        activityIntent.addFlags(Intent.FLAG_ACTIVITY_NEW_TASK);
        //在广播组件里，通过上下文对象启动 Activity 组件
        context.startActivity(activityIntent);
    }
}
```

6.1.3 接收系统短信广播应用实例

Android 设备接收的短信是 PDU（Protocol Description Unit）形式的，短信消息类 SmsMessage 提供了创建该类对象的方法及解析 PDU 格式信息数据的方法，如图 6.1.2 所示。

图 6.1.2 类 SmsMessage 的常用方法

当手机收到短信时，就会产生一个名为 android.provider.Telephony.SMS_RECEIVED 的系统广播，与系统相关的程序不仅会在通知栏里显示该短信，还会把短信的相关信息保存至手机的短信数据库。下面介绍一个处理手机短信的 Android 项目。

【例 6.1.3】 接收处理 Android 手机的短信广播。

广播接收者程序代码如下：

```
public class SmsReceiver extends BroadcastReceiver {
    @Override
    public void onReceive(Context context, Intent intent) {
        //获取发送短信意图对象携带的数据
        Bundle bundle = intent.getExtras();
        if(bundle != null){
            //可使用动态调试，查看短信数据结构
            Object[] pdus = (Object[]) bundle.get("pdus");
            //发送方的一条短信可能被分割为多次发送
            SmsMessage[] msgs = new SmsMessage[pdus.length];
            for (int i = 0; i < msgs.length; i++) {
                byte[] pdu = (byte[]) pdus[i];
                //获取分段的短信
                msgs[i]= SmsMessage.createFromPdu(pdu);
            }
```

```
//构建短信相关信息字符串
StringBuilder strb=new StringBuilder();
for(SmsMessage msg:msgs){
    strb.append("\n 发短信人电话：\n")
        .append(msg.getDisplayOriginatingAddress())
        .append("\n 短信内容：\n")
        .append(msg.getMessageBody());
    //接收时间
    Date date=new Date(msg.getTimestampMillis());
    SimpleDateFormat sdf=new SimpleDateFormat("yyyy-MM-dd HH:mm:ss");
    strb.append("\n 短信接收时间：\n").append(sdf.format(date));
}
//在 context 指示的上下文（就是模块的 MainActivity）里打 Toast 消息
Toast.makeText(context, strb, Toast.LENGTH_LONG).show();
    }
  }
}
```

注意：

（1）在清单文件里，需要注册短信接收权限，在 MainActivity 程序里，需要动态申请短信接收权限。

（2）发送方的一条较长的短信可能被分割为多次发送。

（3）本广播接收者与系统提供的接收者程序并不矛盾，分别起作用。

6.2 自定义广播及其使用

6.2.1 自定义广播

和系统广播不同，自定义广播是由用户创建并用于承载用户自定义信息的广播。前面提到，广播实质上是一个 Intent 类，使用 Intent 对象的 setAction()设置意图动作，调用 Intent 对象的 putExtra()方法传递自定义信息。同时，sendBroadcast()方法和 sendOrderedBroadcast()方法是 Context 类的 2 个成员方法，故 Activity 与 Service 均可调用这两个方法发送自定义广播。

广播消息实质上就是将一个 Intent 对象用 sendBroadcast()方法发送出去。在 Android 系统中，上下文类 Context（为抽象类）提供了发送广播的抽象方法。

在一个 Activity 里，通过使用继承的方法 sendBroadcast()可以实现对自定义广播的发送。

在广播消息前，通过使用 Intent 对象的 putExtras()方法封装广播的数据（为 Bundle 类型），在广播接收程序中通过使用 Intent 对象的 getExtras()方法获得 Bundle 类型的数据。

6.2.2 以动态注册方式使用自定义广播

动态注册是在程序里通过使用 Context 类提供的方法 registerReceiver()完成的。对于操作特别频繁的广播事件（如锁屏和解锁），其广播接收者在清单文件里面注册是无效的，只能在代码里进行动态的注册。

【例 6.2.1】 以动态注册方式使用自定义广播。

程序 MainActivity 的代码如下：

```java
/*
    运行时动态发送广播，自发自收
    发送广播时，可以携带 Bundle 类型的数据供广播接收者对象使用
    抽象类 BroadcastReceiver 包含抽象方法 onReceive()
 */
public class MainActivity extends AppCompatActivity {

    private TextView tv;
    //创建广播接收者对象，在有广播到来时工作（因为没有过滤广播）
    private BroadcastReceiver myReceiver = new BroadcastReceiver() {
        @Override
        public void onReceive(Context context, Intent intent) {
            //获取广播携带的数据，需要清楚发送广播时是如何设置的
            Bundle bundle = intent.getExtras();
            String str1 = bundle.getString("data1");
            String str2 = bundle.getString("data2");
            Toast.makeText(context, str1+" " +str2,Toast.LENGTH_LONG).show();
            tv.setText(str1+" " +str2);
        }
    };

    @Override
    protected void onCreate(Bundle savedInstanceState) {
        super.onCreate(savedInstanceState);
        setContentView(R.layout.activity_main);
        //使用意图过滤器对象
        IntentFilter intentFliter = new IntentFilter();
        //设置接收的广播
        intentFliter.addAction("com.example.broadcast.MY_BROADCAST");
        //动态注册广播接收者
        registerReceiver(myReceiver, intentFliter);
        tv = findViewById(R.id.tv);
        findViewById(R.id.btn).setOnClickListener(new View.OnClickListener() {
            @Override
            public void onClick(View arg0) {
                //创建广播意图对象，广播名称在注册广播接收者使用
                Intent intent = new Intent("com.example.broadcast.MY_BROADCAST");
```

```
                Bundle bundle = new Bundle();
                bundle.putString("data1", "自定义广播与接收案例");
                bundle.putString("data2", "张粤~...0v0...");
                intent.putExtras(bundle);           //捆绑数据
                //发送广播
                sendBroadcast(intent);
                try {
                    Thread.sleep(1000);             //休眠一下，模拟广播可能存在的延迟
                } catch (InterruptedException e) {
                    e.printStackTrace();
                }
            }
        });
    }
}
```

注意：

（1）在程序里创建广播接收者对象，与在清单文件里注册广播接收者作用相同。

（2）广播从发送到接收可能存在一定的延迟。

6.2.3 以静态注册方式使用自定义广播

广播接收者在清单文件中的注册是通过使用标签<receiver>完成的，这种注册方式称为静态注册。

【例 6.2.2】 以静态注册方式实现的自定义广播及接收。

本项目（模块）是例 6.1.1 和例 6.2.1 的混合体，即需要在清单文件中注册广播接收者，而广播是在 Activity 程序里动态发出的，其主要步骤如下。

（1）在项目 Example6 里，新建名为 example6_3a 的模块。

（2）MainActivity 程序的主要功能是动态发送携带了数据的广播，其代码如下：

```
/*
    运行时动态发送广播，静态注册广播接收者
    发送广播时，可以携带 Bundle 类型的数据供广播接收者程序使用
    广播从发送到接收可能存在一定的延迟
    抽象类 BroadcastReceiver 包含抽象方法 onReceive()
*/
public class MainActivity extends AppCompatActivity {
    @Override
    protected void onCreate(Bundle savedInstanceState) {
        super.onCreate(savedInstanceState);
        setContentView(R.layout.activity_main);
        findViewById(R.id.btn).setOnClickListener(new View.OnClickListener() {
            @Override
            public void onClick(View arg0) {
                Intent intent = new Intent("com.example.broadcast.MY_BROADCAST");
                //Android 8.0，可提供更加安全、可靠的服务
```

```
                //下一行在 Android 8.0 中必须，防止用户程序唤醒其他用户应用
                intent.setPackage("com.example.example6_3a");
                Bundle bundle = new Bundle();
                bundle.putString("data1", "自定义广播与接收案例");
                bundle.putString("data2", "张粤~...0v0...");
                intent.putExtras(bundle);          //捆绑数据
                sendBroadcast(intent);             //发送广播
                try {
                    Thread.sleep(1000);            //休眠一下，模拟广播可能存在的延迟
                } catch (InterruptedException e) {
                    e.printStackTrace();
                }
            }
        });
    }
}
```

（3）新建广播接收者程序 MyReceiver，其代码如下：

```
public class MyReceiver extends BroadcastReceiver {
    @Override
    public void onReceive(Context context, Intent intent) {
        Bundle mybundle = intent.getExtras();
        String str1 = mybundle.getString("data1");
        String str2 = mybundle.getString("data2");
        Toast.makeText(context, str1+" " +str2,Toast.LENGTH_LONG).show();
    }
}
```

（4）在清单文件里，注册接收广播的意图过滤器，其代码如下：

```
<receiver
    android:name="com.example.example6_3a.MyReceiver"
    android:enabled="true"
    android:exported="true">
    <intent-filter>
        <action android:name="com.example.broadcast.MY_DROADCAST"/>
    </intent-filter>
</receiver>
```

6.3 通　　知

6.3.1 通知与通知类 Notification

手机通知位于状态栏，在屏幕的最上层，通常显示电池电量、信号强度等信息。按住状态栏往下拉就可以打开查看系统提示信息。

当手机收到短信或来电未接时,都会由应用程序自动产生一个通知。系统时间的数字文本、手机网络连接方式的图标、手机充电指示图标等表示系统特定的状态信息,都在手机的状态栏中显示。

在 Android 中,通知能够通过多种方式提供给用户,如发出声音、设备振动、在位于手机屏幕顶部的状态栏里放置一个持久的图标等。

当手指从状态栏向下滑过时,会展开所有的通知信息,能查看到通知的标题和内容等信息(如未查看的短信),还能清除通知;当手指从下方向上滑过时将隐藏通知。

注意:Android 中有些广播接收者程序对收到的广播进行处理的结果以通知形式展示。

Android 系统提供了对通知进行描述的类 Notification,Android API 16 及以上的版本都推荐使用 Notification 的内部构造器类 Builder 来创建一个 Notification 对象,Notification 类封装了在状态栏里设置发布通知的标题、内容、时间、通知到来时的提示音和标题前的小图标等,如图 6.3.1 所示。

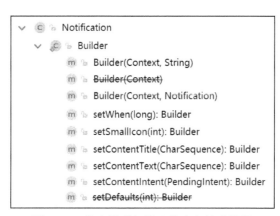

图 6.3.1　状态栏通知类及其内部构造器类

注意:Android 8.0 对内部类 Notification.Builder 的构造方法进行了调整。

6.3.2　通知管理器类 NotificationManager

通知是由 NotificationManager 类型的对象进行管理的,而 NotificationManager 对象需要调用系统服务 Context.NOTIFICATION_SERVICE 来创建,其代码如下:

```
String ns = Context.NOTIFICATION_SERVICE;
NotificationManager nm=(NotificationManager) getSystemService(ns);
```

通知管理器类 NotificationManager 主要提供了发通知、取消通知两个方法。

6.3.3　使用 PendingIntent 查看通知内容

Intent 一般用于 Activity、Service、BroadcastReceiver 之间的数据传递,而 PendingIntent 一般用在 Notification 上。可以理解为,PendingIntent 是延期执行的 Intent,是对 Intent 的一个包装。

【例 6.3.1】 延期意图类 PendingIntent 的使用。

在 MainActivity 程序里，按钮监听器包含了在通知栏创建打开手机设置程序的延期意图对应的通知，也能取消自行发布在通知栏里的指定通知。程序运行时，展开通知栏的效果，如图 6.3.2 所示。

图 6.3.2　程序运行后展开手机通知栏的效果

```
/*
    通过内部类 Notification.Builder 来构建通知（图标、标题和内容）
    (NotificationManager)getSystemService(NOTIFICATION_SERVICE)得到通知管理器对象
    通知管理器对象具有发通知和取消通知的方法
    延期意图 PendingIntent 使得在用户单击通知标题后查看通知内容成为可能
*/
public class MainActivity extends AppCompatActivity {
    NotificationManager notificationManager;    //通知管理器
    Notification.Builder builder;                //通知内部构造器
    Button btn_notification;                     //按钮
    boolean isCreate = false;                    //通知未创建
    @Override
    protected void onCreate(Bundle savedInstanceState) {
        super.onCreate(savedInstanceState);
        setContentView(R.layout.activity_main);
        btn_notification = findViewById(R.id.btn_notification);
        //创建设置意图对象
        Intent intent = new Intent(Settings.ACTION_SETTINGS);
```

```java
        //创建延期意图对象
        PendingIntent pendingIntent = PendingIntent.getActivity(this,0,intent,0);
        //获取通知管理器
        notificationManager = (NotificationManager) getSystemService(NOTIFICATION_SERVICE);
        //使用内部的通知构造器
        builder = new Notification.Builder(this);
        //构建通知内容
        builder.setSmallIcon(R.drawable.notification)        //图标使用矢量图形
                .setContentTitle("设置")
                .setContentText("单击进入设置程序")
                .setWhen(System.currentTimeMillis())
                .setDefaults(Notification.DEFAULT_SOUND)
                .setContentIntent(pendingIntent);            //关键方法
        btn_notification.setOnClickListener(new View.OnClickListener() {
            @Override
            public void onClick(View v) {
                isCreate = !isCreate;                        //在创建与取消之间切换
                TextView tv=findViewById(R.id.tv);
                if(isCreate) {                               //如果准备创建通知
                    //发送通知至通知栏
                    notificationManager.notify(1, builder.build());   //发通知
                    tv.setVisibility(View.VISIBLE);          //设置可见
                    btn_notification.setText("取消通知");
                }else {
                    notificationManager.cancel(1);           //取消通知
                    btn_notification.setText("创建通知");
                    tv.setVisibility(View.INVISIBLE);        //设置不可见
                }
            }
        });
    }
}
```

注意：

（1）Android 8.0 发通知使用新的 API，本程序代码只适合 Android 8.0 以下版本！

（2）在本书配套课程网站上，本章实验给出了兼容 Android 8.0 的通知创建代码。

（3）延期意图类 PendingIntent 的使用（参见第 9 章 NFC 项目）。

习 题 6

一、判断题

1．当手机收到短信时，只产生通知而不产生广播。
2．Android 系统中的所有广播接收者都会对广播进行各自处理。
3．开发短信接收与发送程序时，都要使用短信管理器类 SmsMessage。
4．Notification 与 Toast 一样，出现在手机特定的位置。
5．在 Android Studio 中，BroadcastReceiver 组件可以通过向导的方式来创建。
6．在 Service 组件程序里，也能发送广播。
7．手机通知栏的通知（Notification），用户只能查看，不能编辑。

二、选择题

1．组件 BroadcastReceiver 定义在包_____里。
　　A．android.location　　　　　　　　B．android.app
　　C．android.content　　　　　　　　 D．android.media
2．方法 onReceive()的 2 个参数类型依次是_____。
　　A．Intent 和 Context　　　　　　　　B．Bundle 和 Intent
　　C．Intent 和 Bundle　　　　　　　　 D．Context 和 Intent
3．方法 registerReceiver()参数必须包含_____类型。
　　A．BroadcastReceiver 和 IntentFilter　B．BroadcastReceiver 和 Intent
　　C．Intent 和 Bundle　　　　　　　　 D．IntentFilter 和 Intent
4．类 Notification 定义在包_____里。
　　A．android.app　　　　　　　　　　B．android.widget
　　C．android.content　　　　　　　　 D．android.view
5．通知管理器提供了发通知方法 notify()，其参数个数为_____。
　　A．1　　　　　　B．2　　　　　　C．3　　　　　　D．4

三、填空题

1．在清单文件里注册 BroadcastReceiver 组件，使用_____标签。
2．广播接收者需要重写 BroadcastReceiver 提供的抽象方法_____。
3．在清单文件里注册 BroadcastReceiver 组件时，必须使用标签<intent-flter>并内嵌_____标签，用以指定需要处理的广播。
4．方法 unregisterReceiver()参数的类型是_____。
5．发送或接收的广播数据，由_____类型的对象携带。

第 6 章 广播组件与通知

实 验 6

一、实验目的

（1）掌握常用的系统广播（静态注册）。
（2）掌握获取短信广播数据的编程方法。
（3）掌握自定义广播的发送与接收方法（动态注册）。
（4）掌握通知的使用。

二、实验内容及步骤

【预备】在 Android Studio 中，新建名为 Example6 的项目后，访问本课程配套的网站 http://www.wustwzx.com/as/sy/sy06.html，复制相关代码，完成如下几个模块的设计。

1. 手机短信所产生的系统广播，启动音乐播放程序（参见例 6.1.1）

（1）在项目 Example 6 里，新建名为 example6_1 的模块。
（2）复制一个音乐文件至文件夹 res/raw 里，并命名为 white.mp3。
（3）在程序文件夹里，使用菜单 File→New→Other→BroadcastReceiver，创建名为 SmsReceiver 的广播接收者程序，输入播放音乐的代码。
（4）在清单文件里，注册短信接收权限代码；在意图过滤器代码里，配置广播接收者所处理的广播。
（5）编辑界面程序 MainActivity，增加接收短信的动态权限处理代码，并在 onDestroy() 方法里添加取消广播接收者注册的代码。
（6）安装本模块至手机后，从其他手机向本机发送一条短信，手机接收短信后便开始播放音乐。

2. 手机短信所产生的系统广播，启动播放音乐服务程序（参见例 6.1.2）

（1）在项目 Example 6 里，新建名为 example6_2 的模块。
（2）复制一个音乐文件至文件夹 res/raw 里，并命名为 white.mp3。
（3）在程序文件夹里，使用菜单 File→New→Service，创建名为 MyAudioService 的服务程序，重写组件的相关方法，实现音乐的播放与停止。
（4）编辑界面程序 MainActivity，编写接收短信的动态权限处理代码和停止音乐播放按钮的监听器代码。
（5）在程序文件夹里，创建名为 SMSReceiver 的广播接收者程序，编写启动音乐播放服务和激活 Activity 组件的代码。
（6）在清单文件里，注册短信接收权限代码；在意图过滤器代码里，配置广播接收者所处理的广播。
（7）安装本模块至手机后，从其他手机向本机发送一条短信，手机接收短信后便开始播放音乐，单击按钮可停止音乐播放。

3．获取手机短信内容的程序设计（参见例 6.1.3）

（1）在项目 Example 6 里，新建名为 example6_1a 的模块。

（2）在清单文件里注册短信接收权限，在 MainActivity 程序里动态请求短信接收权限。

（3）在程序文件夹里，创建名为 SMSReceiver 的广播接收者程序，依次编写获取发送短信意图对象携带的数据、使用类 SMSMessage 的相关方法获取短信相关信息的代码后，以 Toast 方式展示解析到的短信息。

（6）安装本模块至手机后，从其他手机向本机发送一条短信，查看测试结果。

4．自定义广播（参见例 6.2.1 和例 6.2.2）

（1）在项目 Example 6 里，新建名为 example6_3 的模块。

（2）在 MainActivity 里，使用类 BroadcastReceiver 的构造方法，创建它的一个实例对象 myReceiver 作为类 MainActivity 的成员（属性），并重写方法 onReceive()。

（3）在 MainActivity 的方法 onCreate()里，创建意图过滤器对象 intentFliter，对其应用方法 addAction()来指定过滤的广播名称，最后使用方法 registerReceiver(myReceiver, intentFliter)完成广播接收者的动态注册。

（4）在方法 onCreate()里，设置发送广播按钮的单击事件监听器，在广播意图对象中使用的广播名称与注册广播接收者指定的名称相同。

（5）安装本模块至手机后做运行测试。

5．通知程序设计（参见例 6.3.1）

（1）在项目 Example 6 里，新建名为 example6_4 的模块。

（2）依次定义类 MainActivity 的四个成员：通知管理器对象 NotificationManager、通知的内部构造器对象 Builder、按钮对象 btn_Notification 和布尔类型的变量 isCreate（初始值为 false）。

（3）在 MainActivity 的方法 onCreate()里，依次创建对象 NotificationManager、Builder 和 btn_Notification。

（4）在 btn_Notification 的监听器代码里，对 NotificationManager 对象应用方法 notify(1,builder.build())发送通知，应用方法 cancel(1)取消通知。

（5）安装本模块至手机后做运行测试。

三、实验小结及思考

（由学生填写，重点写上机中遇到的问题。）

第 7 章　SQLite 数据库编程

在第 4 章，介绍了使用 SharedPreferences 和文件存储数据。当频繁地存储数据时，需要使用数据库来管理，数据库一直是 Android 应用的重要组成部分。在 Android 系统中，通常使用适合于嵌入式系统的 SQLite 数据库，因为它占用资源少、运行高效、可移植性好，并且提供了零配置的运行模式。Android 系统使用 SQLite 存储、管理、维护数据，对数据库的创建、打开、更新和记录（增加、删除、修改、查询）是数据库的基本内容。本章学习要点如下：

- 掌握 SQLite 数据库的创建、打开和更新方法；
- 使用 SQLiteDatabase 类实现对 SQLite 数据库增加、删除、修改和查询的操作；
- 掌握 SQLiteDatabase 类提供的用于数据库增加、删除、修改和查询的操作的多种方法；
- 掌握以 DAO 方式编写数据库访问程序的方法；
- 掌握数据库事务的使用方法。

7.1　SQLite 数据库简介

7.1.1　SQLite 数据库软件的特点

SQLite 是 Android 系统手机自带（内置）的轻量级数据库软件，类似于 Access 数据库软件，它提供了对数据库增加、删除、修改和查询的操作。此外，它还具有如下特点：

- 占用资源少，特别适合于嵌入式设备；
- 支持较多的开发语言；
- 无需安装和配置；
- 使用关系型的 SQL 命令进行 SQLite 数据库操作。

7.1.2　Android 系统对 SQLite 数据库的支持

在标准的 Android 软件包里，提供了与 SQLite 数据库相关的两个软件包。一个包是 android.database，它定义了游标接口 Cursor（与 JDBC 中的 ResultSet 类似）；另一个包是 android.database.sqlite，它包含了 2 个抽象类 SQLiteOpenHelper 和 SQLiteDatabase，如图 7.1.1 所示。

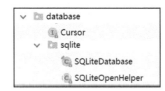

图 7.1.1 Android 提供的 SQLite 软件包及其相关类

7.2 使用 SQLiteOpenHelper 创建、打开或更新数据库

7.2.1 SQLite 数据库及表的创建与打开

抽象类 SQLiteOpenHelper 位于包 android.database.sqlite 内，定义了子类必须重写的 2 个方法 onCreate()和 onUpgrade()，不仅提供了创建、打开数据库操作的构造方法 SQLiteOpenHelper()，还提供了获取 SQLiteDatabase 对象的方法（getReadableDatabase() 和 getWritableDatabase()），如图 7.2.1 所示。

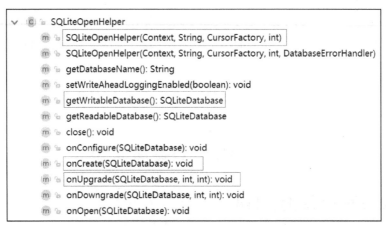

图 7.2.1 使用抽象类 SQLiteOpenHelper 的主要方法

注意：使用抽象类 SQLiteOpenHelper 的构造方法创建数据库时，以上下文对象作为其中的 1 个参数。

由于 Java 不支持多继承，为了代码复用，通常在访问 SQLite 数据库的项目里，建立抽象类 SQLiteOpenHelper 的子类作为一个独立类，或者作为 Activity 组件程序的内部类，其示例代码如下：

```java
public class MyDbHelper extends SQLiteOpenHelper {
    publicMyDbHelper(Context context) {
        //super(context, name, factory, version); //调用父类构造方法
        //第 2 个参数为库名，第 4 个参数为库版本号
        super(context, "test.db", null, 1);
    }
```

第 7 章　SQLite 数据库编程

```
@Override
public void onCreate(SQLiteDatabase sqLiteDatabase) {
    //操作数据库的 SQL 命令，在创建新数据库时自动执行
}
@Override
public void onUpgrade(SQLiteDatabase sqLiteDatabase, int i, int i1) {
    //操作数据库的 SQL 命令，在数据库版本提升时自动执行
}
}
```

使用 SQLiteOpenHelper 的方法 getWritableDatabase()可获得 SQLiteDatabase 类型的数据库对象。类 SQLiteDatabase 封装了对数据库操作的方法，如图 7.2.2 所示。

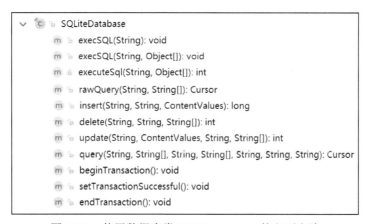

图 7.2.2　使用数据库类 SQLiteDatabase 的主要方法

7.2.2　使用 SQLiteSpy 验证创建的数据库

Android 应用创建的数据库位于文件夹 data/data/包名/databases 里，使用 Device File Explorer 找到应用创建的 SQLite 数据库，在数据库名的菜单里，选择 Save As 选项，即可将该 SQLite 数据库导入计算机，其操作如图 7.2.3 所示。

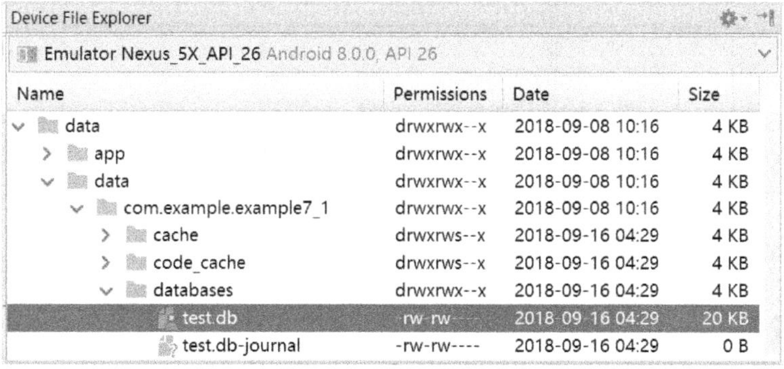

图 7.2.3　找到 AVD 里的 SQLite 数据库

SQLiteSpy 是一款在 Windows 中运行、以图形方式管理 SQLite 数据库的软件，在编

者的教学网站里提供了下载链接，解压后即可使用（无须安装），其运行界面如图 7.2.4 所示。

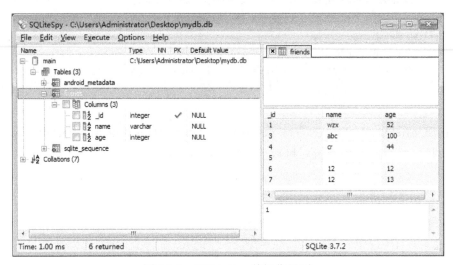

图 7.2.4　SQLiteSpy 的操作界面

7.2.3　SQLite 数据库的更新

SQLiteDatabase 类提供了对数据库操作查询的 execSQL(SQL)，其中 SQL 参数是对数据库执行操作查询的 SQL 命令（而不是使用 Select 命令的普通查询），通常是 Insert、Delete、Update 或 Create Table、Alter Table 等命令。

【例 7.2.1】　SQLite 数据库及表的创建与更新。

【程序运行】程序首次部署到手机运行后，创建数据库 test.db 及表 person，其中 person 表包含两个字段。当修改源程序，将数据库版本从 1 提升到 2 时，再次部署、运行程序，会执行 onUpgrade()方法，修改 person 表结构，增加一个字段。导出数据库，并使用 SQLiteSpy 可查验。

程序 MainActivity 的结构，如图 7.2.5 所示。

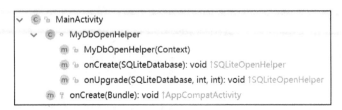

图 7.2.5　MainActivity 类结构图

【实现步骤】

编写工程的主界面程序 MainActivity.java，其源代码如下：

```
/*
    首次安装时，调用 SQLiteOpenHelper 的 onCreate()方法创建库 test.db 及表 person
```

```
    以后运行时，若数据库没有版本提升，则用可读写方式直接打开数据库；
    若有版本提升，则先执行 onUpgrade()方法后，再用可读写方式打开数据库
*/
public class MainActivity extends AppCompatActivity {
    @Override
    protected void onCreate(Bundle savedInstanceState) {
        super.onCreate(savedInstanceState);
        setContentView(R.layout.activity_main);
        MyDbOpenHelper helper = new MyDbOpenHelper(this);         //调用构造方法创建数据库
        SQLiteDatabase db=helper.getWritableDatabase();           //可读写
        //操作数据库的代码
    }
    class MyDbOpenHelper extends SQLiteOpenHelper {               //内部工具类
        public MyDbOpenHelper(Context context) {
            //必须调用抽象父类的构造方法，第4个参数为数据库版本
            super(context, "test.db", null,1);                    //建立或打开库
        }
        @Override
        public void onCreate(SQLiteDatabase db) {
            Toast.makeText(MainActivity.this, "创建数据库表...", Toast.LENGTH_SHORT). show();
            db.execSQL("CREATE TABLE person(id Integer Primary Key AutoIncrement,
                                            name VARCHAR(20))");
            db.execSQL("insert into person values(null,'Wu')");
            db.execSQL("insert into person values(null,'Guan')");
        }
        @Override
        public void onUpgrade(SQLiteDatabase db, int oldVersion, int newVersion) {
            Toast.makeText(MainActivity.this, "数据库表更新中...", Toast.LENGTH_SHORT).show();
            db.execSQL("ALTER TABLE person ADD tel CHAR(20)");
            db.execSQL("update person set tel='15527643858' where name='Wu'");
            db.execSQL("update person set tel='1340862750' where name='Guan'");
        }
    }
}
```

7.3 使用 SQLiteDatabase 实现数据库表的增加、删除、修改和查询

7.3.1 记录的增加、删除、修改和查询

1. 操作数据库的两种方式

操作数据库有两种方式，一种是使用方法 execSQL()实现记录的增加、删除和修改，使用方法 rawQuery()查询数据库表，它们都以一条完整的 SQL 命令作为方法参数。

注意：方法 execSQL()可以实现数据库表结构的创建和修改。

操作数据库的另一种方式是使用方法 insert()、update()、delete()和 query()实现记录的增加、删除、修改和查询，这些方法的参数不是完整的 SQL 命令，而是 SQL 命令中的相关短语。其中，方法 insert()和 update()使用了 ContentValues 类型的参数。

注意：

（1）execSQL(String,Object[])方法较 execSQL(String)而言，在 SQL 命令中支持通配符 "?"。

（2）方法 insert()、update()、delete()和 query()中的第 1 个参数都是表名。

（3）执行 insert()等方法，本质上也是执行 SQL 命令，只是系统已经组装相关参数为完整的 SQL 语句。

2．使用接口 Cursor 表示查询结果

使用 SQL 命令查询数据库得到的是内存中的一张虚拟表（记录集），通过记录指针可以遍历所有记录，游标接口 android.database.Cursor 就是为这个目的设计的。

含有两个参数的原生查询方法 rawQuery()，它的第 1 个参数是完整的 SQL 命令，第 2 个参数是条件参数，配合 SQL 命令使用的占位符 "?"。例如：

```
Cursor c=db.rawQuery("select * from person where id=?",new String[]{idv+""});
```

注意：由于 id 字段是整型，因此上面的 idv 表达式应为整型。作为 SQL 命令的一部分，它需要转换成字符串。比较标准的做法是使用 "Integer(idv).toString()"。

含 7 个参数的 query()方法，各参数的含义在 Android Studio 中有提示，第 1 个参数为表名，第 2 个参数为字段集，第 3 个参数为选择条件，第 4 个参数为条件参数，最后一个参数为排序依据。例如，查询表 person 中年龄低于 50 岁的人员代码如下：

```
String selectFilter="age<50";
Cursor c=db.query("person", null, selectFilter, null, null, null, "_id ASC");
```

注意：较原生查询方法 rawQuery()而言，query()这种非原生查询方式更加灵活。

SQLiteDatabase 类的两种查询方法都是得到 Cursor 接口类型的对象，该接口提供的常用方法如图 7.3.1 所示。

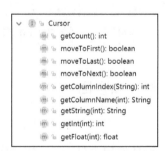

图 7.3.1　Cursor 接口提供的常用方法

3. 使用 ContentValues 类型的参数

SQLiteDatabase 的方法 insert()和 update()都包含了 ContentValues 类型的参数，类 ContentValues 提供了"键—值"对形式的数据，其主要方法如图 7.3.2 所示。

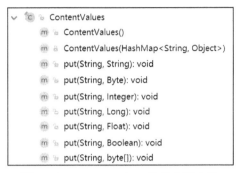

图 7.3.2 ContentValues 类的主要方法

例如，插入记录常用方法为 insert(table,null,contentValues)。

注意：当 contentValues 参数为空或者里面没有内容的时候，insert()会失败，因为底层数据库不允许增加一个空行。为了防止这种情况，可以将第 2 个参数设置为一个列名。如果能保证 contentValues 值非空，则可以将第 2 个参数简单地设置为 "null"。

7.3.2 使用适配器 SimpleAdapter 显示查询结果

在第 3.3.6 小节，介绍了简单的 ArrayAdapter 数据适配器。为了在 ListView 等控件中显示查询得到的结果集，完成一条记录的多项输出，可以使用 SimpleAdapter 数据适配器来完成。

数据适配器类 SimpleAdapter 提供的构造方法用于定义适配器。除此之外，它还有许多方法，如图 7.3.3 所示。

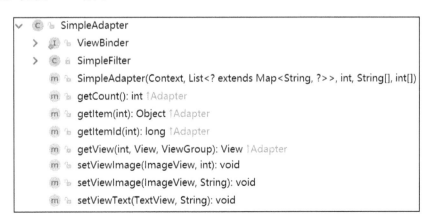

图 7.3.3 数据适配器类 SimpleAdapter

定义数据适配器 SimpleAdapter 时，共需要 5 个参数，分别是上下文对象、泛型列表数据、列表布局文件名（不带扩展名）、列表数据中的字段名数组和列表布局项控件名称数组。定义与使用 SimpleAdapter 的参考代码如下：

```
private Cursor cursor;
ListView listView;
private Map<String,Object> listItem;              //列表项，对应一条记录
private List<Map<String,Object>> listData;        //所有列表数据
//查询数据库并赋值给 listData
while(cursor.moveToNext()){
    listItem=new HashMap<String,Object>();        //创建实例
    //分别获取各字段值给相应变量：id、name 和 age
    listItem.put("_id",id);                       //键值对
    listItem.put("name",name);
    listItem.put("age",age);
    listData.add(listItem);
}
//创建数据适配器对象并填充数据
SimpleAdapter listAdapter = new SimpleAdapter(
    this,
    listData,
    R.layout.list_item,
    //列表数据中的字段集，对应于 Map 集合里的键名
    new String[]{"_id","name","age"},
    //列表布局中用于显示字段数据的控件集
    new int[]{R.id.tv_id,R.id.tv_name,R.id.tv_age});
listView.setAdapter(listAdapter);                 //绑定数据显示控件
```

注意：

（1）由于 SimpleAdapter 的一个列表项里包含了多项数据，因此，在使用 ArrayList 定义其数据项之前，需要使用 HashMap 定义其列表项；

（2）创建数据适配器 SimpleAdapter 对象时，其构造方法中共包含 5 个参数；

（3）通过 ListView 等控件提供的 setAdapter()方法实现数据的绑定。

7.3.3 以 DAO 方式访问数据库编写程序

如果将数据库访问的 CRUD 方法都写在 Activity 组件，会导致结构不清晰。使用 DAO（Database Access Object）方式可以将数据库操作的业务代码封装到一个类文件的相关方法里，其具有如下优点：

（1）代码便于重用；

（2）代码冗余度低；

（3）结构更清晰。

【例 7.3.1】 使用 DAO 方式，访问 SQLite 数据库。

项目运行时，向数据库表增加 2 条记录并使用 ListView 控件显示效果，单击按钮，可以实现增加、修改和删除功能，如图 7.3.4 所示。

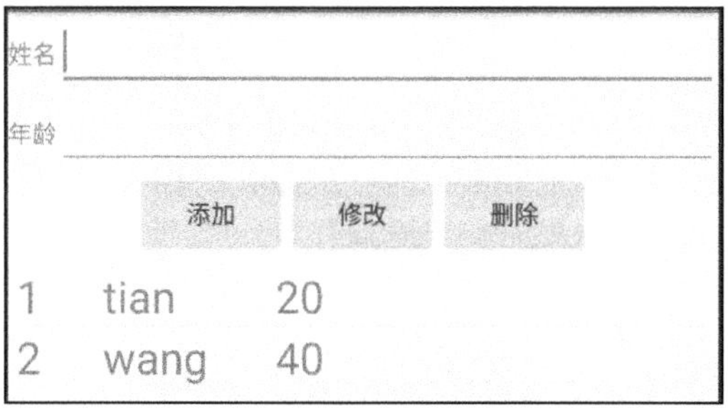

第 7 章　SQLite 数据库编程

图 7.3.4　项目运行效果

主要设计步骤如下。

(1) 抽象类 SQLiteOpenHelper 的子类 MyDbHelper，其代码如下：

```
public class MyDbHelper extends SQLiteOpenHelper{
    public static final String TB_NAME = "friends";        //表名
    //构造方法：第 1 个参数为上下文，第 2 个参数为数据库名，第 3 个参数为游标工厂，第 4
                                                          个参数为版本
    public MyDbHelper(Context context, String dbname, CursorFactory factory, int version) {
        super(context, dbname, factory, version);           //创建或打开数据库
    }
    @Override
    public void onCreate(SQLiteDatabase db) {
        //当表不存在时，创建表；第一字段为自增长类型
        db.execSQL("CREATE TABLE IF NOT EXISTS " +
                TB_NAME + "( _id integer primary key autoincrement," +
                "name varchar," + "age integer"+ ")");
    }
    @Override
    public void onUpgrade(SQLiteDatabase db, int oldVersion, int newVersion) {
        // 执行 SQL 命令
        db.execSQL("DROP TABLE IF EXISTS " + TB_NAME);
        onCreate(db);
    }
}
```

(2) 封装 CRUD 方法到类文件 MyDAO.java，其代码如下：

```
/*
    本类 MyDAO 调用了打开数据库的助手类 MyDbHelper
    本类 MyDAO 提供的 CRUD 针对数据库 test.db 的表 friends
    查询数据库表所有记录的方法：allQuery()
    插入记录的方法：insertInfo(String name,int age)
    删除记录的方法：deleteInfo(String selId)
    修改记录的方法：updateInfo(String name,int age,String selId)
*/
public class MyDAO {
```

```java
    private SQLiteDatabase myDb;                           //类的成员
    private MyDbHelper dbHelper;                           //类的成员
    public MyDAO(Context context) {                        //构造方法,参数为上下文对象
        //第1个参数为上下文,第2个参数为数据库名
        dbHelper = new MyDbHelper(context,"test.db",null,1);
        myDb = dbHelper.getReadableDatabase();
    }
    public Cursor allQuery(){                              //查询所有记录
        return myDb.rawQuery("select * from friends",null);
    }
    public int getRecordsNumber(){                         //返回数据表记录数
        Cursor cursor = myDb.rawQuery("select * from friends",null);
        return cursor.getCount();
    }
    public void insertInfo(String name,int age){           //插入记录
        myDb = dbHelper.getWritableDatabase();
        ContentValues values = new ContentValues();
        values.put("name", name);
        values.put("age", age);
        long rowid=myDb.insert(MyDbHelper.TB_NAME, null, values);
        if(rowid==-1)
            Log.i("myDbDemo", "数据插入失败! ");
        else
            Log.i("myDbDemo", "数据插入成功! "+rowid);
    }
    public void deleteInfo(String selId){                  //删除记录
            String where = "_id=" + selId;
            int i = myDb.delete(MyDbHelper.TB_NAME, where, null);
            if (i > 0)
                Log.i("myDbDemo", "数据删除成功! ");
            else
                Log.i("myDbDemo", "数据未删除! ");
    }
    public void updateInfo(String name,int age,String selId){    //修改记录
        //方法中的第3个参数用于修改选定的记录
        ContentValues values = new ContentValues();
        values.put("name", name);
        values.put("age", age);
        String where="_id="+selId;
        int i=myDb.update(MyDbHelper.TB_NAME, values, where, null);
        //上面几行代码的功能可以用下面的一行代码实现
        //myDb.execSQL("update friends set name = ? ,age = ? where _id = ?",
                                                 new Object[]{name,age,selId});
        if(i>0)
            Log.i("myDbDemo","数据更新成功! ");
        else
```

```
            Log.i("myDbDemo","数据未更新！");
        }
    }
}
```

(3) 编写 MainActivity 程序，其代码如下：

```
/*
    本程序中对数据库的插入操作和查询，使用了 MyDAO 类的相关方法
    首次运行时，增加两条记录并使用 ListView 控件显示出来
*/
public class MainActivity extends AppCompatActivity implements View.OnClickListener{
    private MyDAO myDAO;                    //数据库访问对象
    private ListView listView;
    private List<Map<String,Object>> listData;
    private Map<String,Object> listItem;
    private SimpleAdapter listAdapter;
    private EditText et_name;               //数据表包含 3 个字段，第 1 字段为自增长类型
    private EditText et_age;
    private   String selId=null;            //选择项 Id
    @Override
    protected void onCreate(Bundle savedInstanceState) {
        super.onCreate(savedInstanceState);
        setContentView(R.layout.activity_main);
        Button bt_add= (Button) findViewById(R.id.bt_add);bt_add.setOnClickListener(this);
        Button bt_modify=(Button)findViewById(R.id.bt_modify);bt_modify.setOnClickListener (this);
        Button bt_del=(Button)findViewById(R.id.bt_del);bt_del.setOnClickListener(this);
        et_name=(EditText)findViewById(R.id.et_name);
        et_age=(EditText)findViewById(R.id.et_age);
        myDAO = new MyDAO(this);            //创建数据库访问对象
        if(myDAO.getRecordsNumber()==0) {    //防止重复运行时插入记录
            myDAO.insertInfo("tian", 20);    //插入记录
            myDAO.insertInfo("wang", 40);    //插入记录
        }
        displayRecords();                    //显示记录
    }
    public void displayRecords(){            //显示记录方法定义
        listView = (ListView)findViewById(R.id.listView);
        listData = new ArrayList<Map<String,Object>>();
        Cursor cursor = myDAO.allQuery();
        while (cursor.moveToNext()){
            int id=cursor.getInt(0);                //获取字段值
            String name=cursor.getString(1);
            //int age=cursor.getInt(2);
            int age=cursor.getInt(cursor.getColumnIndex("age"));    //推荐此种方式
            listItem=new HashMap<String,Object>();              //必须在循环体里新建
            listItem.put("_id", id);                //第 1 个参数为键名，第 2 个参数为键值
```

```java
            listItem.put("name", name);
            listItem.put("age", age);
            listData.add(listItem);                    //添加一条记录
        }
        listAdapter = new SimpleAdapter(this,
                listData,
                R.layout.list_item,
                new String[]{"_id","name","age"},
                new int[]{R.id.tv_id,R.id.tvname,R.id.tvage});
        listView.setAdapter(listAdapter);              //应用适配器
        listView.setOnItemClickListener(new AdapterView.OnItemClickListener() {    //列表项监听
            @Override
            public void onItemClick(AdapterView<?> parent, View view, int position, long id) {
                //从适配器读取记录
                Map<String,Object> rec= (Map<String, Object>) listAdapter.getItem (position);
                et_name.setText(rec.get("name").toString());         //刷新文本框
                et_age.setText(rec.get("age").toString());
                Log.i("ly",rec.get("_id").toString());
                selId=rec.get("_id").toString();        //供修改和删除时使用
            }
        });
    }
    @Override
    public void onClick(View v) {                     //实现的接口方法
        if(selId!=null) {                             //选择了列表项后,可以增加、删除、修改
            String p1 = et_name.getText().toString().trim();
            int p2 = Integer.parseInt(et_age.getText().toString());
            switch (v.getId()){
                case    R.id.bt_add:
                    myDAO.insertInfo(p1,p2);
                    break;
                case    R.id.bt_modify:
                    myDAO.updateInfo(p1,p2,selId);
                    Toast.makeText(getApplicationContext(),"更新成功! ",
                            Toast.LENGTH_SHORT).show();
                    break;
                case    R.id.bt_del:
                    myDAO.deleteInfo(selId);
                    Toast.makeText(getApplicationContext(),"删除成功! ",
                            Toast.LENGTH_SHORT).show();
                    et_name.setText(null);et_age.setText(null); selId=null;    //提示
            }
        }else{                                        //未选择列表项
            if(v.getId()==R.id.bt_add) {              //单击"添加"按钮
                String p1 = et_name.getText().toString();
                String p2 = et_age.getText().toString();
```

```
            if(p1.equals("")||p2.equals("")){                    //要求输入信息
                Toast.makeText(getApplicationContext(),"姓名和年龄都不能空！",
                                        Toast.LENGTH_SHORT).show();
            }else{
                myDAO.insertInfo(p1, Integer.parseInt(p2));    //第 2 个参数转型
            }
        } else{                                              //单击修改或删除按钮
            Toast.makeText(getApplicationContext(),"请先选择记录！",
                                        Toast.LENGTH_SHORT).show();
        }
    }
    displayRecords();                                        //刷新 ListView 对象
}
```

由于在 DAO 类方法中的参数很多是数据表的字段，因此，可以将各个数据表对应地建立一个实体类，这些类通常称为 Model 层。每个实体类除了包含那些字段成员外，只包含 set×××()和 get×××()方法。在 DAO 类的方法中使用实体类对象参数，减少了 DAO 类方法中参数的个数，使程序更加清晰。

使用"DAO+实体类+BaseAdapter"方式，访问 SQLite 数据库。

与例 7.3.1 相比，添加了实体类，Activity 组件使用适配器 BaseAdapter，而类 MyDbHelper 与 MyDAO 相同。

（1）实体类 Person.java 的代码如下：

```
public class Person {
    private int _id;                                         //数据库里是自增长字段
    private String name;
    private int age;
    Person(int _id,String name,int age){
        this._id=_id;
        this.name=name;
        this.age=age;
    }
    public void set_id(int _id) {
        this._id = _id;
    }
    public void setName(String name) {
        this.name = name;
    }
    public void setAge(int age) {
        this.age = age;
    }
    public int get_id() {
        return _id;
    }
```

```java
    public String getName() {
        return name;
    }
    public int getAge() {
        return age;
    }
}
```

（2）MainActivity.java 代码如下：

```java
public class MainActivity extends AppCompatActivity implements View.OnClickListener{
    MyDAO myDAO;                          //数据库访问对象
    ListView listView;                    //列表控件
    List<Person> listData;                //列表数据
    BaseAdapter adapter;                  //列表控件的数据适合器
    //数据表包含 3 个字段，第 1 字段为自增长类型
    int selId;                            //选择项 Id，与自增长类型的第 1 字段相对应
    EditText et_name;                     //对应第 2 字段的控件对象
    EditText et_age;                      //对应第 3 字段的控件对象
    @Override
    protected void onCreate(Bundle savedInstanceState) {
        super.onCreate(savedInstanceState);
        setContentView(R.layout.activity_main);
        //3 个按钮
        Button bt_add= findViewById(R.id.bt_add);bt_add.setOnClickListener(this);
        Button bt_modify=findViewById(R.id.bt_modify);bt_modify.setOnClickListener(this);
        Button bt_del=findViewById(R.id.bt_del);bt_del.setOnClickListener(this);
        //两个编辑文本框
        et_name=findViewById(R.id.et_name);
        et_age=findViewById(R.id.et_age);
        //初次安装时建立两条记录
        myDAO = new MyDAO(this);             //创建数据库访问对象
        if(myDAO.getRecordsNumber()==0) {    //防止重复运行时插入记录
            myDAO.insertInfo("tian", 20);    //插入记录
            myDAO.insertInfo("wang", 40);    //插入记录
        }
        displayRecords();                    //显示记录
    }
    public void displayRecords(){            //定义显示记录方法
        listView = findViewById(R.id.listView);
        listData = new ArrayList<Person>();
        Cursor cursor = myDAO.allQuery();
        while (cursor.moveToNext()){         //遍历记录集
            int id=cursor.getInt(0);         //获取字段值
            String name=cursor.getString(1);
            int age=cursor.getInt(cursor.getColumnIndex("age"));  //推荐此种方式
            Person person=new Person(id,name,age);                //创建 Person 对象
```

```java
            listData.add(person);                                    //添加一条记录
        }
        adapter=new BaseAdapter() {                      //创建适配器，需要重写4个抽象方法
            @Override
            public int getCount() {
                return listData.size();
            }
            @Override
            public Object getItem(int position) {
                return listData.get(position);
            }
            @Override
            public long getItemId(int position) {
                return position;
            }
            @Override
            public View getView(int position, View convertView, ViewGroup parent) {
                // 根据布局文件创建列表项 View（不同于 SimpleAdapter）
                View item = View.inflate(getApplicationContext(), R.layout.list_item, null);
                // 获取这个动态生成列表项中的各个字段对应的文本标签
                TextView idTV = item.findViewById(R.id.tv_id);
                TextView nameTV =   item.findViewById(R.id.tvname);
                TextView ageTV = item.findViewById(R.id.tvage);
                // 根据位置获取 Person 对象（列表项数据）
                Person p = listData.get(position);
                /*设置标签文本（不同于 SimpleAdapter）。其中，_id 和 age 两个字段是整型,
                不做字符串处理时，将导致闪退。用到实体类的 get/set 访问方法*/
                idTV.setText(p.get_id()+"");
                nameTV.setText(p.getName());
                ageTV.setText(p.getAge()+"");
                return item;
            }
        };
        listView.setAdapter(adapter);                              //应用适配器
        listView.setOnItemClickListener(new AdapterView.OnItemClickListener() {  //列表项监听
            @Override
            public void onItemClick(AdapterView<?> parent, View view, int position, long id) {
                Person rec= (Person) adapter.getItem(position);     //从适配器读取记录
                et_name.setText(rec.getName());                     //刷新文本框
                et_age.setText(rec.getAge()+"");
                selId=rec.get_id();                                 //供修改和删除时使用
            }
        });
    }
    @Override
    public void onClick(View v) {                        //实现的接口方法
```

```java
            if(selId>0) {                                    //选择了列表项后,可以增加、删除、修改
                String p1 = et_name.getText().toString().trim();
                int p2 = Integer.parseInt(et_age.getText().toString());
                switch (v.getId()){
                    case    R.id.bt_add:
                        myDAO.insertInfo(p1,p2);
                        break;
                    case    R.id.bt_modify:
                        myDAO.updateInfo(p1,p2,selId);
                        Toast.makeText(getApplicationContext(),"更新成功!",
                                                    Toast.LENGTH_SHORT).show();
                        break;
                    case    R.id.bt_del:
                        myDAO.deleteInfo(selId);
                        Toast.makeText(getApplicationContext(),"删除成功!",
                                                    Toast.LENGTH_SHORT).show();
                        et_name.setText(null);et_age.setText(null); selId=0;   //提示
                }
            }else{                                           //未选择列表项
                if(v.getId()==R.id.bt_add) {                 //单击"添加"按钮
                    String p1 = et_name.getText().toString();
                    String p2 = et_age.getText().toString();
                    if(p1.equals("")||p2.equals("")){        //要求输入信息
                        Toast.makeText(getApplicationContext(),"姓名和年龄都不能空!",
                                                    Toast.LENGTH_SHORT).show();
                    }else{
                        myDAO.insertInfo(p1, Integer.parseInt(p2));   //第2个参数转型
                    }
                } else{                                      //单击"修改"或"删除"按钮
                    Toast.makeText(getApplicationContext(),"请先选择记录!",
                                                    Toast.LENGTH_SHORT).show();
                }
            }
            displayRecords();                                //刷新 ListView 对象
        }
    }
```

7.3.4 使用数据库事务

使用数据库事务的典型案例就是银行转账汇款。当一个账户减少金额时,另一个账户增加金额,而不能出现一个账户减少金额后系统抛出异常,导致另一个账户没有增加金额,反之亦然。处理这类同时提交或同时不提交的业务,就是数据库事务。

SQLiteDatabase 提供了处理数据库事务的相关方法。例如,开启数据库事务的 beginTransaction()方法、处理完毕数据库事务的 setTransactionSuccessful()方法、结束数据库事务的 endTransaction()方法等。

【例7.3.2】 使用数据库事务,实现银行汇款。

初次运行程序,会产生除数为0的异常,账户余额不会变化,账户111的余额为2000,账户222的余额为8000,导出数据库可查验。屏蔽程序中会产生异常的代码,再次部署和运行程序,则交易成功,账户余额同时变化,账户111的余额为1500,账户222的余额为8500,导出数据库可查验。

程序代码如下:

```java
public class MainActivity extends AppCompatActivity {
    MyDbOpenHelper helper;
    SQLiteDatabase db;
    @Override
    protected void onCreate(Bundle savedInstanceState) {
        super.onCreate(savedInstanceState);
        setContentView(R.layout.activity_main);
        //创建或连接数据库
        helper = new MyDbOpenHelper(this);
        db = helper.getWritableDatabase();
        setMit();                        //执行自定义的事务处理方法
        db.close();
    }
    public   void setMit(){              //事务处理方法
        db.beginTransaction();           //数据库事务
        try{
            int num = 500;               //汇款金额
            String from = "111";         //汇出账户
            String to = "222";           //接收账户
            db.execSQL("update accounts set balance = balance - ? where id = ?",
                    new Object[]{num,from});
            //产生除数为0的异常,屏蔽此行才能完成事务
            int temp=1/0;
            db.execSQL("update accounts set balance = balance + ? where id = ?",new
                    Object[]{num,to});
            db.setTransactionSuccessful();//事务成功标识
            Toast.makeText(this, "交易成功", Toast.LENGTH_LONG).show();
        } catch (Exception e){
            Toast.makeText(MainActivity.this, "异常中断,交易未能成功",
                    Toast.LENGTH_SHORT).show();
        } finally{
            db.endTransaction();
        }
    }
    class MyDbOpenHelper extends SQLiteOpenHelper {   //自定义工具内部类
        public MyDbOpenHelper(Context context) {
            super(context, "test.db", null,1);        //建立或打开库
        }
```

```java
            @Override
            public void onCreate(SQLiteDatabase db) {
                //创建表结构
                db.execSQL("CREATE TABLE accounts(id char(6) ,name VARCHAR(6), balance double)");
                //插入2条记录
                db.execSQL("insert into accounts values('111', 'tian',2000)");
                db.execSQL("insert into accounts values('222', 'liu',8000)");
            }
            @Override
            public void onUpgrade(SQLiteDatabase db, int oldVersion, int newVersion) {
                db.execSQL("DROP TABLE IF EXISTS accounts");
                onCreate(db);
            }
        }
    }
```

习 题 7

一、判断题

1．应用程序创建的 SQLite 数据库为该应用私有。
2．SQLiteOpenHelper 的 2 个方法 onCreate()和 onUpgrade()会同时执行。
3．接口 Cursor 与抽象类 SQLiteOpenHelper 所处的软件包相同。
4．使用数据库事务能保证一组代码同时执行。
5．Android 应用创建的数据库，不卸载而重新安装该应用时，不会被删除。

二、选择题

1．使用 SQLiteOpenHelper 的构造方法创建与数据库的连接时，其参数可以设置为 null 的是_____。
    A．上下文   B．数据库名
    C．游标工厂   D．数据库版本

2．SQLiteDatabase 提供的_____方法以完整的 SQL 命令作为参数。
    A．insert()   B．query()
    C．update()   D．execSQL()

3．含有 SQLite 数据库访问的 Android 应用中，其数据库存放在应用包名文件夹的_____文件夹里。
    A．databases   B．lib   C．files   D．shared_prefs

4．数据适配器 SimpleAdapter 的构造函数中参数个数是_____。
    A．3   B．4   C．5   D．6

5．设 Cursor 对象 cursor 的第 3 个字段 age 为 int 类型，则访问该字段的正确方法是_____。
    A．int age=cursor.getString(2);
    B．int age=cursor.getString(3);
    C．int age=cursor.getInt(3);
    D．int age=cursor.getInt(cursor.getColumnIndex("age"));

6．数据库事务的相关方法封装在_____类中。
    A．SQLiteOpenHelper   B．SQLiteDatabase
    C．ContentValues   D．Cursor

三、填空题

1．SQLite 数据库文件的扩展名为_____。
2．在 Android 系统中，创建与连接 SQLite 数据库需要使用抽象类_____。

3．在 Android 的_____类中，封装了对 SQLite 数据库及表的操作方法。
4．SQLiteDatabase 提供 execSQL()方法的返回值类型是_____。
5．当程序提升了数据库版本，重新部署后，会执行_____方法。
6．游标接口 Cursor 提供 moveToNext()方法的返回类型是_____。

实 验 7

一、实验目的

（1）掌握 SQLite 数据库的建立、打开和更新方法。
（2）掌握对数据库中表结构增加、删除、修改的方法。
（3）掌握对数据库中表内容增加、删除、修改和查询的方法。
（4）掌握数据适配器 SimpleAdapter 和 BaseAdapter 的使用。
（5）掌握 DAO 方式在数据库中的应用。
（6）掌握数据库事务的使用方法。

二、实验内容及步骤

【预备】在 Android Studio 中，新建名为 Example7 的项目后，访问本课程配套的网站 http://www.wustwzx.com/as/sy/sy07.html，复制相关代码，完成如下几个模块的设计。

1. 使用 SQLiteOpenHelper 和 SQLiteDatabase 创建、更新数据库及表（参见例 7.2.1）

（1）在项目 Example 7 里，新建名为 example7_1 的模块。

（2）在 MainActivity 里，创建一个继承抽象类 SQLiteOpenHelper 的助手类 MyDbHelper 作为内部类成员。在其构造方法里，调用父类构造方法，在当前上下文环境里指向一个数据库并指定其版本。

（3）重写 MyDbHelper 父类的两个抽象方法 onCreate()和 onUpgrade()，编写数据库的操作代码，如创建表、插入记录等。

（4）在 MainActivity 的 onCreate()方法里，创建类 MyDbOpenHelper 的实例，调用其方法 getWritableDatabase()获得一个数据库句柄。

（5）部署应用到 Android 设备。使用 Android Studio 的设备文件浏览器，从文件夹 data/data/包名/databases 里导出已经创建的数据库至计算机桌面。

（6）从本课程配套网站主页的 Android 版块里，下载 SQLite 数据库操作软件 SQLiteSpy，压缩后直接运行（无须安装），打开刚才导出的数据库。查看对应于类 MyDbHelper 的 onCreate()方法代码的数据库操作情况。

（7）在对应于类 MyDbHelper 的构造方法里，提高数据库的版本后再次部署（不卸载）模块、导出数据库，可查验其不会执行类 MyDbHelper 的 onCreate()方法而是执行了 onUpgrade()方法。

2. 以"DAO+SimpleAdapter"方式访问数据库（参见例 7.3.1）

（1）在项目 Example 7 里，新建名为 example7_2 的模块。
（2）在程序包名里，创建独立类 MyDbHelper（不是内部类），其内容与上相同。

（3）在程序包名里，创建类 MyDAO，分别定义 SQLiteDatabase 和 MyDbHelper 类型的成员 myDb 和 dbHelper。在其构造方法里，创建类 MyDbHelper 的实例 dbHelper，进而得到 myDb 对象。

（4）在 MyDAO 里，封装访问数据库的 CRUD 方法。

（5）编写 MainActivity 程序，使用适配器 SimpleAdapter 在 ListView 控件上展示二维表数据。

（6）在布局文件里，定义 3 个按钮，以使用增加、修改和删除功能。

（7）部署模块、做运行测试。

3．以"DAO+实体类+BaseAdapter"方式访问数据库（参见例 7.3.2）。

（1）在项目 Example 7 里，新建名为 example7_2a 的模块。

（2）复制模块 example7_2 里的类 MyDbHelper 和 MyDAO 至本模块的程序文件夹。

（3）在程序包名里，编写实体类 Person。

（4）在 MainActivity 程序里，列表数据定义为 List<Person>类型。

（5）使用 BaseAdapter 作为 ListView 控件的数据适配器。

（6）部署模块、做运行测试。

4．数据库事务的使用（参见例 7.3.3）。

（1）在项目 Example 7 里，新建名为 example7_3 的模块。

（2）在 MainActivity 程序里，创建内部类 MyDbHelper。在 MyDbOpenHelper 的 onCreate()方法内创建一个包含两个账户的表。

（3）在 MainActivity 的 onCreate()方法开启数据库事务，一个操作是一个账户转出资金，另一个账户是转入资金，两个操作之间安排一个分母为 0 的异常操作。

（4）部署和运行模块，使用 Android Studio 的设备文件浏览器导出数据库并使用 SQLiteSpy 打开，观察数据库事务没有成功执行（两个账户的资金没有变化）。

（5）屏蔽分母为 0 的那行代码，再次重复上面的操作，可观察到数据库事务成功执行（2 个账户的资金有变化）。

三、实验小结及思考

（由学生填写，重点写上机中遇到的问题。）

第 8 章 Android 内容提供者组件

在前面介绍的应用中,一个应用程序创建的数据对于另一个应用程序是不可见的。实际上,也可以将 Android 应用创建的数据暴露到外界,供其他应用程序使用,这样可以达到应用程序间的数据共享,减少数据冗余。之前的文件操作模式(参见文件存储和 SharedPreferences 存储相关内容),通过指定 Context.MODE_WORLD_READABLE 或 Context.MODE_WORLD_WRITEABLE 也可以实现对外共享数据。但是,因为数据存储方式的不同,无法使用统一的数据访问方式。

ContentProvider 是实现应用程序间数据共享最标准的方式,也是介绍的 Android 四大组件中的最后一个。应用程序通过 ContentResolver 对象访问 ContentProvider 中的数据,该对象提供了持久层数据的 CRUD 方法。本章学习要点如下:
- 掌握使用 ContentProvider 建立内容提供者的方法;
- 掌握通过 ContentResolver 访问 ContentProvider 的方法;
- 掌握 Android 提供的常用公共数据接口;
- 掌握手机联系人数据库的结构信息;
- 通过读取手机联系人的应用,进一步掌握 UI 设计和数据适配器的使用。

8.1 ContentProvider 组件及其相关类

8.1.1 抽象类 ContentProvider(内容提供者)

作为 Android 的四大组件之一,内容提供者 android.content.ContentProvider 是 Android 系统中不同应用程序之间共享数据的接口,用于保存和检索数据。

ContentProvider 是 Android 提供给上层的一个组件,主要用于实现数据访问的统一管理和数据共享。这里的数据管理是通过定义统一的访问接口来完成,如增加、删除、修改、查询。同时,它采用了类似 Internet 的 URL 机制,将数据以 URI 形式标识,这样其他 App 就可以采用一套标准的 URI 规范来访问同一处数据,而不用关心具体的实现细节。在 Android 系统中可能会涉及一个 App 的数据被其他 App 使用的情况,比如通讯录、日历、短信等,这时就需要一套能实现数据共享的机制,这里的 ContentProvider 就可以提供该功能,其底层使用了 Binder 来完成 App 进程之间的通信,同时使用匿名共享内存来作为共享数据的载体。当然为了保证数据访问的安全性，ContentProvider 还对每处的数据 URI 增加了权限管理机制，以控制该数据的访问者及访问方式。

ContentProvider 作为一个数据源处理中心，不止给 App 内部使用，还会被其他 App 访问。所以这里的 ContentProvider 就好比一个 server，其他的 App 在访问这个 ContentProvider 的时候都必须先获得一个 Server 的 Client 才行，这里的 ContentResolver 就是 Android 提供给开发者使用得到一个 ContentProvider Client 的管理工具类。

使用 ContentProvider 访问某个应用程序的数据，可以不必关心某个应用程序的数据存储方式是使用数据库方式还是文件方式，还是通过网络获取数据，这些都不重要，重要的是其他应用程序可以通过一个 ContentProvider 类型的对象来操作某个应用程序的数据。

Android 应用开发者将自己的持久化数据（采用 File 存储或 SQLite 数据库存储，通常为后者）公开给其他应用程序，可以有两种方法：一是定义自己的 ContentProvider 子类，二是将当前应用程序的数据添加到已有的 ContentProvider 中。

新建一个继承于 ContentProvider 的类 TestContentProvider 时，自动创建源文件 TestContentProvider.java 的代码如下：

```java
import android.content.ContentProvider;        //主类
import android.content.ContentValues;          //相关类
import android.database.Cursor;                //相关类
import android.net.Uri;                        //相关类
public class TestContentProvider extends ContentProvider {
    @Override
    public boolean onCreate() {
        return false;
    }
    @Nullable
    @Override
    public Cursor query(@NonNull Uri uri, @Nullable String[] projection, @Nullable String
                        selection, @Nullable String[] selectionArgs, @Nullable String sortOrder) {
        return null;
    }
    @Nullable
    @Override
    public String getType(@NonNull Uri uri) {
        return null;
    }
    @Nullable
    @Override
    public Uri insert(@NonNull Uri uri, @Nullable ContentValues values) {
        return null;
    }
    @Override
```

```
    public int delete(@NonNull Uri uri, @Nullable String selection,
                                          @Nullable String[] selectionArgs) {
        return 0;
    }
    @Override
    public int update(@NonNull Uri uri, @Nullable ContentValues values,
                      @Nullable String selection, @Nullable String[] selectionArgs) {
        return 0;
    }
}
```

Android 的抽象类 ContentProvider，其定义如图 8.1.1 所示。

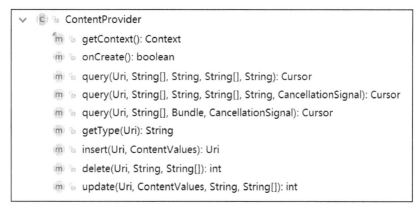

图 8.1.1 抽象类 ContentProvider 的定义

注意：组件 ContentProvider 的 4 个 CRUD 方法中第 1 个参数均为 Uri 类型。

8.1.2 抽象类 ContentResolver（内容解析器）

每个 Activity 都有一个 ContentResolver 对象，并且通过 Activity 的超类 Context（表示上下文）的 getContentResolver()方法获得该对象，抽象类 Context 的定义，如图 8.1.2 所示。

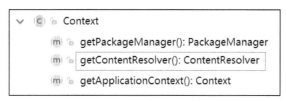

图 8.1.2 抽象类 Context 的定义

android.content.ContentResolver 是与 ContentProvider 相关的一个类，称为内容解析器。通过 ContentResolver 对象访问 ContentProvider 中的数据时，系统要求把 ContentProvider 封装成一个 URI。

抽象类 ContentResolver 也提供了持久层数据的 CRUD 方法，如图 8.1.3 所示。

图 8.1.3 抽象类 ContentResolver 的定义

其中，CRUD 方法中的第 1 个参数也为 Uri 类型，与 ContentProvider 的方法相对应。

注意：

（1）ContentResolver 与 ContenProvider 具有相同的 CRUD 方法。

（2）ContentResolver 的方法 insert(Uri url, ContentValues values)将一条记录的数据插入到 Uri 指定的地方，并返回最新添加那个记录的 Uri。

8.1.3 内容提供者的 Uri 定义及其相关类（UriMatcher 和 ContentUris）

在第 4.2.1 小节，类 android.net.Uri 是用来表示通用资源的。作为内容提供者，还需要提供其他应用访问的路径。为了表示内容提供者这种资源，Android 规定了它的 scheme 为"content"。

Android 系统附带的 ContentProvider 有如下几种。

- Browser：存储浏览器的信息；
- CallLog：存储通话记录等信息；
- Contacts：存储手机联系人等信息；
- MediaStore：存储媒体文件的信息；
- Settings：存储设备的设置和首选项信息。

Android 2.0 以上版本提供了访问手机联系人信息的多种 URL。例如，访问电话信息 Uri 的常量写法为：

ContactsContract.CommonDataKinds.Phone.CONTENT_URI

为了实现对多张表的灵活访问，通常调用传入的 URI 时，在 ContentProvider 中使用 android.content.UriMatcher 类来解析 URI 地址，即使用工具类 UriMatcher 定义匹配规则。工具类 UriMatcher 的定义如图 8.1.4 所示。

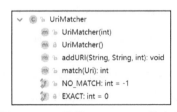

图 8.1.4 工具类 UriMatcher 的定义

在内容提供者中，使用 UriMatcher 来定义匹配规则的示例代码如下：

```
UriMatcher matcher=new UriMatcher(UriMatcher.NO_MATCH);
//下面的 com.example.contentprovider.myprovider 为内容提供者字符串
//匹配到 person 时返回 1
matcher.addURI("com.example.contentprovider.myprovider","person/#",1);
//匹配到 student 时返回 2
matcher.addURI("com.example.contentprovider.myprovider","student",2);
DbHelper helper=new DbHelper(this.getContext());
SQLiteDatabase db=helper.getReadableDatabase();
//匹配从使用内容提供者程序中传入的 Uri
switch(matcher.match(uri)) {
    case 1:
        return db.query("person",projection, selection,selectionArgs,null,null,sortOrder);
    case 2:
        return db.query("student ",projection, selection,selectionArgs,null,null,sortOrder);
    default:
        throw new RuntimeException("Uri 不能识别："+uri);
}
```

定义了上面的匹配规则后，在使用内容提供者的程序中，就可以使用如下的 URI：

Uri uri=Uri.parse("content:// com.example.contentprovider.myprovider/person/10");

注意：

（1）在对 UriMatcher 对象注册时，第 1 个参数为主机名（Authority），第 2 个参数为路径名，可以理解为要操作的数据库中表的名字，第 3 个参数为匹配码（匹配不成功时，返回全部数据）；

（2）在匹配 URI 时，可以使用 "#" 代表任意数字，使用 "*" 代表任意文本。

android.content.ContentUris 也是一个辅助类，其定义如图 8.1.5 所示。

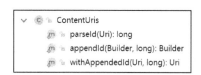

图 8.1.5　工具类 ContentUris 的定义

方法 withAppendedId()把 URI 和 ID 连接成一个新的 URI，这在内容提供者程序的插入时将用到追加记录方法。

ContentUris 类提供的方法 parseId()负责把 URI 后边的 ID 解析出来。

8.2　自定义 ContentProvider 及其使用

8.2.1　在 Android 应用里创建并注册内容提供者

Android 系统支持任意应用程序自己创建的 ContentProvider，以便将本应用程序中的数据提供给其他应用程序,其方法是:在当前应用程序中定义继承 ContentProvider 的子类,并实现 CRUD。

如同使用其他的 Android 组件一样,建立自己的 ContentProvider,需要在系统清单文件里注册,其方法是使用<provider>标签注册,其代码如下:

```
<provider android:name="MyDbProvider"
          android:authorities="introduction.android.mydbdemo.myfriendsdb
          android:exported="true" />
```

其中:name 为必填属性,表示 ContentProvider 子类的名称;authorities 也是必填属性,表示其他应用程序访问该 ContentProvider 时的路径,相当于访问 Web 服务器的域名,原则上可以任意定义,但一般还是使用 ContentProvider 子类的存放路径。

注意:

(1)清单文件中的<provider>标签还有一个任选属性 permission,用于对本应用提供数据访问的限定。当省略 permission 属性时,表示任何应用都可以访问本应用提供的数据。

(2)项目部署后,测试内容提供者 URI 发布成功的代码如下:

```
ContentResolver contentResolver = getContentResolver();
Uri uri = Uri.parse("content://com.example.mydbprovider.MYPROVIDER");  //适当修改字符串
Log.i("测试","内容提供者记录数:"+contentResolver.query(uri,null,null,null,null).getColumnCount());
```

【例 8.2.1】 发布内容提供者。

(1)在项目 Example 8 里,新建名为 example8_2 的模块。

(2)在程序文件夹里,右击包名→New→Other→Content Provider,输入内容提供组件名称 MyDbProvider 及应用的 Uri。此时,内容提供者组件会在清单文件里注册。

(3)在程序文件夹里,新建名为 MyDAO 的数据库访问类,按照内容提供者组件定义的方法参数和返回值类型设计,其代码如下:

```
/*
    本类供设计 ContentProvider 组件时使用
    使用第 7 章已经定义过的数据库助手类 DbHelper
*/
public class MyDAO {
    private DbHelper dbHelper;          //数据库助手
    private SQLiteDatabase myDb;        //数据库对象
    //内容提供者的 uri
    private Uri uri= Uri.parse("content://com.example.mydbprovider.MYPROVIDER");
    public MyDAO(Context context) {     //构造方法,参数 context 为上下文对象
        dbHelper = new DbHelper(context,"test.db",null,1);  //第 1 个参数为上下文,第 2 个参数
                                                             为数据库名
        myDb = dbHelper.getWritableDatabase();
    }
    //下面的 4 个方法,与内容提供者的 CRUD 方法相对应(注意返回值类型)
```

```java
    public Uri insertInfo(ContentValues values){                        //插入记录
        long rowid=myDb.insert(DbHelper.TB_NAME, null, values);
        if(rowid==-1) {
            Log.i("myDbDemo", "数据插入失败！");
            return null;
        } else{
            Log.i("myDbDemo", "数据插入成功！"+rowid);
            Uri insertUri = ContentUris.withAppendedId(uri,rowid);
            return insertUri;
        }
    }
    //下面的3个方法，并未使用内容提供者的 Uri
    public int deleteInfo(Uri uri, String selection, String[] selectionArgs){      //删除记录
        //使用表名代替形参数 uri
        return myDb.delete(DbHelper.TB_NAME,selection, selectionArgs);
    }
    public int updateInfo(Uri uri, ContentValues values, String selection, String[] selectionArgs){
                                                                        //修改记录
        //使用表名代替形参数 uri
        return myDb.update(DbHelper.TB_NAME, values, selection,selectionArgs);
    }
    public Cursor allQuery(){                                           //查询所有记录方法：
        myDb = dbHelper.getReadableDatabase();
        return myDb.rawQuery("select * from friends",null);
    }
}
```

（4）使用类 MyDAO 的相关方法，编写内容提供者组件的相应方法，其代码如下：

```java
/*
    调用符合 ContentProvider 规范的 MyDAO 类（不同于 example7_2 模块里的 MyDAO）
    MyDbProvider 在清单文件里注册为 com.example.mydbprovider.MYPROVIDER
*/
public class MyDbProvider extends ContentProvider {
    private Context context;
    private MyDAO myDAO;
    @Override
    public boolean onCreate() {
        context = this.getContext();
        myDAO = new MyDAO(context);
        return true;
    }
    @Override
    public String getType(Uri uri) {
        throw new UnsupportedOperationException("Not yet implemented");
    }
    @Override
```

```
        public Cursor query(Uri uri, String[] projection, String selection, String[] selectionArgs, String
                                                                                           sortOrder) {
            return myDAO.allQuery();
        }
        @Override
        public Uri insert(Uri uri, ContentValues values) {
            return myDAO.insertInfo(values);
        }
        @Override
        public int delete(Uri uri, String selection, String[] selectionArgs) {
            return myDAO.deleteInfo(uri,selection,selectionArgs);
        }
        @Override
        public int update(Uri uri, ContentValues values, String selection, String[] selectionArgs) {
            return myDAO.updateInfo(uri,values,selection,selectionArgs);
        }
    }
```

（5）部署模块至手机，内容提供者组件的 URI 发布成功。

8.2.2　在另一个应用程序里使用内容提供者

在另一个应用程序里，使用内容提供者的依据是已经在清单文件中注册 ContentProvider 提供的访问域名，可以在当前应用程序中使用该 ContentProvider。

【例 8.2.2】 使用已经发布的内容提供者。

【运行测试】运行例 8.2.1 的内容提供者程序，显示两条记录，退出再运行本内容提供者使用程序，增加一条记录后退出，再次运行内容提供者程序，可查验已经增加了一条记录，即两个程序操作相同的数据源。

（1）在项目 Example 8 里，新建名为 example8_2a 的模块。

（2）程序 MainActivity.java 的代码如下：

```
public class MainActivity extends AppCompatActivity implements View.OnClickListener{
    EditText et_name;
    EditText et_age;
    Button bt_add,bt_modlfy,bt_del;
    ListView listView;
    SimpleAdapter listAdapter;
    ArrayList<Map<String,Object>> listData;
    Map<String,Object> listItem;
    String selID;
    ContentResolver contentResolver;
    Uri uri = Uri.parse("content://com.example.mydbprovider.MYPROVIDER");
    private Cursor cursor = null;
    @Override
    protected void onCreate(Bundle savedInstanceState) {
        super.onCreate(savedInstanceState);
```

```java
setContentView(R.layout.activity_main);
Toast.makeText(this, "使用内容提供者模块", Toast.LENGTH_SHORT).show();
et_name = findViewById(R.id.et_name);
et_age=   findViewById(R.id.et_age);
listView = findViewById(R.id.listView);
listData = new ArrayList<Map<String, Object>>();
bt_add = findViewById(R.id.bt_add);bt_add.setOnClickListener(this);
bt_modify = findViewById(R.id.bt_modify);bt_modify.setOnClickListener(this);
bt_del = findViewById(R.id.bt_del);bt_del.setOnClickListener(this);
//获取内容解析器
contentResolver = getContentResolver();
dbFindAll();
//列表项监听
listView.setOnItemClickListener(new AdapterView.OnItemClickListener() {
    @Override
    public void onItemClick(AdapterView<?> parent, View view, int position, long id) {
        Map<String,Object> listItem = (Map<String, Object>) listView. getItemAtPosition
                                                                    (position);
        et_name.setText((CharSequence) listItem.get("name"));
        et_age.setText((CharSequence) listItem.get("age"));
        selID= (String) listItem.get("_id");
        Toast.makeText(MainActivity.this, "id:"+selID, Toast.LENGTH_SHORT). show();
    }
});
}
private void dbFindAll() {
    listData.clear();
    cursor=contentResolver.query(uri,null,null,null,null);
    listItem = new HashMap<String,Object>();
    listItem.put("_id","序号");
    listItem.put("name","姓名");
    listItem.put("age","年龄");
    listData.add(listItem);
    if(cursor==null){
        Toast.makeText(this, "未查找到任何数据，请检查 uri 是否正确...",
                                                Toast.LENGTH_SHORT).show();
    }else{
        cursor.moveToFirst();
        while (!cursor.isAfterLast()){
            String id = cursor.getString(0);
            String name = cursor.getString(1);
            String age = cursor.getString(2);
            listItem = new HashMap<String,Object>();
            listItem.put("_id",id);
            listItem.put("name",name);
            listItem.put("age",age);
```

```java
                    listData.add(listItem);
                    cursor.moveToNext();
                }
            }
            showList();
        }
        private void showList() {
            listAdapter = new SimpleAdapter(this,
                    listData,
                    R.layout.list_item,
                    new String[]{"_id","name","age"},
                    new int[]{R.id.tv_id,R.id.tvname, R.id.tvage});
            listView.setAdapter(listAdapter);
        }
        @Override
        public void onClick(View v) {                    //监听器接口方法
            int se=v.getId();
            switch (se){
                case R.id.bt_add:
                    ContentValues values = new ContentValues();
                    values.put("name",et_name.getText().toString().trim());
                    values.put("age",et_age.getText().toString().trim());
                    Uri insertUri = contentResolver.insert(uri,values);
                    Toast.makeText(this, insertUri.toString(), Toast.LENGTH_SHORT).show();
                    dbFindAll();
                    break;
                case R.id.bt_modify:
                    String where ="_id="+selID;
                    ContentValues cv = new ContentValues();
                    cv.put("name",et_name.getText().toString().trim());
                    cv.put("age",et_age.getText().toString().trim());
                    int i = contentResolver.update(uri,cv,where,null);
                    if (i > 0) {
                        Toast.makeText(this, "修改成功!", Toast.LENGTH_SHORT).show();
                    }else {
                        Toast.makeText(this, "修改失败!", Toast.LENGTH_SHORT).show();
                    }
                    dbFindAll();
                    break;
                case R.id.bt_del:
                    String tj = "_id=" + selID;
                    int num = contentResolver.delete(uri, tj, null);
                    if (num > 0) {
                        Toast.makeText(this, "删除成功!", Toast.LENGTH_SHORT).show();
                    }else {
                        Toast.makeText(this, "删除失败!", Toast.LENGTH_SHORT).show();
```

```
            }
            dbFindAll();showList();        //刷新
        }
    }
}
```

注意：使用 ContentProvider 的 Uri 必须与创建 ContentProvider 时指定的 Uri 一致，否则，不会获取到数据。

8.3 读取手机联系人信息

8.3.1 手机联系人相关类 ContactsContract

android.provider.ContactsContract 是关于手机联系人信息的类，其内部类 Contacts 提供了访问手机联系人数据库的 ContactsContract.Contacts.CONTENT_URI。通过这个 Uri，可以访问手机中所有人的相关信息。

8.3.2 手机联系人数据库及其相关表

在手机系统目录 data/data 的包 com.android.providers.contacts 里，存放手机联系人数据库 contacts2.db。

注意：手机取得 Root 权限后，才能查看系统数据库 contacts2.db。

为了满足手机用户个性化设置的需求，如动态地创建手机联系人分组、存放多种联系方式（手机、固定电话）等，手机联系人数据库中包含了多张表而不是一张表。手机联系人数据库较好地解决了手机联系人数据的扩展管理问题。

1．存放手机联系人 ID 的表 raw_contacts

在存放了两个手机联系人数据库 contacts2.db 里，表 raw_contacts 的记录信息如图 8.3.1 所示。

_id	contact_id	pinned	display_name
1	1	2147483647	WuZhiXiang
2	2	2147483647	CaoRong

图 8.3.1 一个存放了 2 个手机联系人的 raw_contacts 表

其中，手机联系人 ID 存放在字段 contact_id 里。

2．手机联系人数据表 Data

在存放了两个手机联系人数据库 contacts2.db 里，表 Data 的记录信息如图 8.3.2 所示。

_id	mimety...	raw_contact_id	data1	data2	data3	data4
1	5	1	1 552-764-3858	2		
2	8	1	02751012663	1		02751012663
3	1	1	707348355@qq.com	1		
4	7	1	WuZhixiang	WuZhixiang		
5	5	2	1 599-422-0815	2		
6	7	2	CaoRong	CaoRong		

图 8.3.2　一个存放了两个手机联系人的 Data 表

注意：在查询手机联系人信息时，需要先查询表 raw_contacts，取得手机联系人的 ID。

3．手机联系人表 contacts

在存放了两个手机联系人数据库 contacts2.db 里，表 contacts 的记录信息如图 8.3.3 所示。

_id	name_raw_contact_id	photo_id	photo_file_id	custom_ringtone	times_contacted	has_phone_number
1	1				0	1
2	2				0	1

图 8.3.3　一个存放了两个手机联系人的 contacts 表

注意：图 8.3.3 所示的表里，字段 _id 中存放手机联系人的 ID，字段 has_phone_number 中存放手机联系人的电话个数。此外，还有存放照片的字段 photo_id 和 photo_file_id。

4．保存数据类型的表 MimeType

表 MimeType 存放了 11 种数据类型，如图 8.3.4 所示。

_id	mimetype
1	vnd.android.cursor.item/email_v2
2	vnd.android.cursor.item/im
3	vnd.android.cursor.item/nickname
4	vnd.android.cursor.item/organization
5	vnd.android.cursor.item/phone_v2
6	vnd.android.cursor.item/sip_address
7	vnd.android.cursor.item/name
8	vnd.android.cursor.item/postal-address_v2
9	vnd.android.cursor.item/identity
10	vnd.android.cursor.item/photo
11	vnd.android.cursor.item/group_membership

图 8.3.4　表 MimeType 存放的数据类型

从图 8.3.4 中可以看出，E-mail、电话和姓名这三种数据分别用 1、5 和 7 表示。

8.3.3　读取手机联系人程序设计

Android 系统对一系列的公用数据类型提供了对应的 ContentProvider 接口。要访问手机联系人，不必直接对系统的 contacts.db 数据库进行操作，可以通过如下方式获取手机联系人信息 Uri：

```
Uri uri = ContactsContract.Contacts.CONTENT_URI;
```

【例 8.3.1】 使用系统提供的公用数据接口，读取并显示手机联系人的姓名及电话。在清单文件中注册读取手机联系人信息的权限，其代码如下：

```
<uses-permission android:name="android.permission.READ_CONTACTS"/>
```

由于手机联系人权限是危险权限，因此，需要在文件 MainActivity.java 里申请读取手机联系人的动态权限。获取 ContentResolver 对象后，根据系统提供的信息读取手机联系人的 URI，查询所有联系人信息。最后，显示到 TextView 控件，MainActivity.java 的代码如下：

```java
/*
    获取手机联系人的 URI，相当于提供了一个公共的数据库链接
    本程序读取手机联系人姓名及电话，需要读联系人权限 android.permission.READ_CONTACTS
    使用 Android 自带的列表布局 android.R.layout.simple_list_item_2
    对 ListView 控件应用 SimpleAdapter 适配器
*/
public class MainActivity extends AppCompatActivity {
    Uri uri = ContactsContract.Contacts.CONTENT_URI;    //内部类静态成员
    ContentResolver contentResolver;
    Cursor cursor;
    ListView listView;
    SimpleAdapter listAdapter;
    HashMap<String, String> item;                       //存放记录数据
    ArrayList<Map<String, String>> data;                //所有记录
    @Override
    public void onCreate(Bundle savedInstanceState) {
        super.onCreate(savedInstanceState);
        setContentView(R.layout.activity_main);
        if (ContextCompat.checkSelfPermission(this, Manifest.permission.READ_CONTACTS) !=
                                        PackageManager.PERMISSION_GRANTED) {
            ActivityCompat.requestPermissions(MainActivity.this,
                            new String[]{Manifest.permission.READ_CONTACTS}, 1);
        } else {
            fun();
        }
    }
    @Override
    public void onRequestPermissionsResult(int requestCode, @NonNull String[] permissions,
                                        @NonNull int[] grantResults) {
        super.onRequestPermissionsResult(requestCode, permissions, grantResults);
        if (grantResults.length > 0) {
            if (grantResults[0] == PackageManager.PERMISSION_GRANTED) {
                fun();
```

```
        } else {
            Toast.makeText(this, "权限不足！", Toast.LENGTH_SHORT).show();
            finish();        //主动销毁当前Activity（不同于用户按返回键）
        }
    }
}
public void fun() {
    //android.content.Context 提供了抽象类 getContentResolver()方法
    //通过内容（数据）解析器使用抽象类 android.content.ContentResolver 提供了 query()等
                                                                                    方法
    cursor = getContentResolver().query(uri, null, null, null, null);      //得到记录集
    listView = findViewById(R.id.listView);
    data = new ArrayList<Map<String, String>>();
    TextView textView = findViewById(R.id.textView);
    textView.setText("读取到" + cursor.getCount() + "个联系人");
    //textView.setText("读取到"+String.valueOf(cursor.getCount())+"个联系人");
    while (cursor.moveToNext()) {
        //先获取联系人_id 字段的索引号后再获取_id 值
        int idFieldIndex = cursor.getColumnIndex("_id");//法一
        //int idFieldIndex=cursor.getColumnIndex(ContactsContract.Contacts._ID); //法二
        int id = cursor.getInt(idFieldIndex);
        //先获取联系人姓名字段的索引号后再获取姓名字段值
        int nameFieldIndex = cursor.getColumnIndex("display_name");
        //int nameFieldIndex=cursor.getColumnIndex(ContactsContract.Contacts.DISPLAY_
                                                                                    NAME);
        String name = cursor.getString(nameFieldIndex);
        int numCountFieldIndex = cursor.getColumnIndex(ContactsContract.
                                                    Contacts.HAS_PHONE_NUMBER);
        int numCount = cursor.getInt(numCountFieldIndex); //获取联系人的电话号码个数
        String phoneNumber = "";
        if (numCount > 0) {                           //联系人有至少一个电话号码
            //在类ContactsContract.CommonDataKinds.Phone中根据id查询相应联系人的所
                                                                                有电话
            Cursor phonecursor = getContentResolver().query(
                    ContactsContract.CommonDataKinds.Phone.CONTENT_URI,
                    null,
                    ContactsContract.CommonDataKinds.Phone.CONTACT_ID + "=?",
                    new String[]{Integer.toString(id)}, null);
            if (phonecursor.moveToFirst()) {          //仅读取第一个电话号码
                int numFieldIndex = phonecursor.getColumnIndex(ContactsContract.
                                                    CommonDataKinds.Phone.NUMBER);
                phoneNumber = phonecursor.getString(numFieldIndex);
            }
        }
        item = new HashMap<String, String>();        //必须循环创建
        item.put("name", name);
        item.put("phoneNumber", phoneNumber);
```

```
            data.add(item);
        }
        listAdapter = new SimpleAdapter(this, data,
                android.R.layout.simple_list_item_2,        //使用 Android 自带的列表布局
                new String[]{"name", "phoneNumber"},
                new int[]{android.R.id.text1, android.R.id.text2});    //数据适配器
        listView.setAdapter(listAdapter);                              //显示
    }
}
```

8.4 Android 后台线程与 Android 组件的综合应用

8.4.1 Android UI 主线程

在 Android 四大组件中，Activity、Service 和 BroadcastReceiver 都工作在主线程上，因此，任何耗时的操作都会降低用户界面的响应速度，甚至导致用户界面失去响应。

一个 Android 程序默认情况下也只有一个进程，在这个进程里可以有许多个线程，其中只有一个称之为 UI 线程。UI 线程在 Android 应用程序运行时被创建，是一个进程中的主线程，主要是负责控制 UI 界面的显示、更新和控件交互等。

在 Android 程序创建之初，一个进程呈现的是单线程模型，所有的任务都在一个线程中运行。因此，UI Thread 所执行的每一个函数，所花费的时间应该是越短越好。而其他比较费时的工作，如访问网络、下载数据和查询数据库等，都应该交给子线程去执行，以免阻塞主线程。

一个 Activity 的各个生命周期是运行在一个应用进程的主线程当中，其代码被运行有两种入口，第 1 种是生命周期的回调方法，如 onCreate()和 onResume()等；第 2 种是 UI 主线程的消息队列分派消息到 Handler 中，由 Handler 的 run()方法在主线程中执行。

在 Android 应用中，如果有 5s 的主线程堵塞系统就会强退。另外，UI 的更新也只能在主线程中完成。因此，异步处理是不可避免的。Android 提供了类 Handler 和 AsyncTask 用于异步处理，也就是为了不阻塞主线程。

注意：如果在子线程中直接更新 UI，则会抛出异常。

8.4.2 使用 Handler 向 UI 线程传递消息

Handler 运行在 UI 主线程中，它与子线程可以通过 Message 对象来传递数据，Handler 负责接受子线程传过携带数据的 Message 对象，并把这些消息放入主线程队列中，配合主线程进行 UI 更新。

类 Handler 及其相关类 Looper、MessageQueen 和 Message 的作用如下：

- Handler：接收和处理的消息对象，可以有多个；
- Looper：读取 MessageQueen 中的消息，读到消息后交给 Handler 去处理，只有 1 个；

- MessageQueen：存放消息的队列；
- Message：传递信息的消息体，为封装数据。

一般地，在主线程中实例化一个 Handler 并实现其内部接口的回调方法（通常用于 UI 更新），在子线程中调用其发送消息 sendMessage()方法。

Handler 及处理消息类的定义如图 8.4.1 所示。

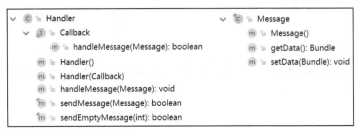

图 8.4.1　Handler 与 Message 类的定义

注意：

（1）Looper 和 MessageQueue 是由 Android 自动管理的。

（2）异步处理本质上使用了 Java 多线程。

【例 8.4.1】　使用 Handler 对象向 UI 线程发送消息。

程序 MainActivity.java 代码如下：

```java
/*
    使用 Handler 对象在后台子线程里向主线程传递消息
    Handler 对象在定义时就设置了消息处理方法
*/
public class MainActivity extends AppCompatActivity {
    Button mButton = null;                                      //按钮
    TextView mTextView = null;                                  //文本框
    Handler mHandler = new Handler() {                          //创建 Handler 对象
        @Override
        public void handleMessage(Message msg) {                //重写类方法
            //super.handleMessage(msg);
            Bundle bundle = msg.getData();
            if (bundle != null) {
                mTextView.setText(bundle.getCharSequence("result"));   //更新 UI 主线程
            } else {
                Toast.makeText(MainActivity.this, "未获得数据！", Toast.LENGTH_SHORT).
                        show();
            }
        }
    };
    /*
    //创建 Handler 对象的方式，实现构造方法，其参数为内部接口
    Handler mHandler = new Handler(new Handler.Callback() {
```

```
            @Override
            public boolean handleMessage(Message msg) {
                Bundle bundle = msg.getData();
                if (bundle != null) {
                    mTextView.setText(bundle.getCharSequence("result"));     //更新 UI 主线程
                } else {
                    Toast.makeText(MainActivity.this, "未获得数据！", Toast.LENGTH_SHORT).
                                                                                   show();
                }
                return false;            //不传递给 Handler 的 handleMessage()方法执行
            }
        });*/
    @Override
    protected void onCreate(Bundle savedInstanceState) {
        super.onCreate(savedInstanceState);
        setContentView(R.layout.activity_main);
        mButton = findViewById(R.id.mButton);
        mTextView = findViewById(R.id.mTextView);
        mButton.setOnClickListener(new View.OnClickListener() {
            @Override
            public void onClick(View view) {
                new Thread(new Runnable() {                      //创建子线程并运行
                    @Override
                    public void run() {
                        //执行下面的代码时，程序将自动闪退
                        /*Toast.makeText(MainActivity.this, "子线程运行中...",
                                                    Toast.LENGTH_SHORT).show();
                        mTextView.setText("这是更新后的信息");//更新 UI 主线程*/
                        Message message = new Message();
                        Bundle bundle = new Bundle();
                        bundle.putCharSequence("result", "这是更新后的数据。");
                        message.setData(bundle);
                        mHandler.sendMessage(message);           //发送消息方法
                    }
                }).start();
            }
        });
    }
}
```

注意：使用 Handler 发送消息代码只能在子线程里。

8.4.3 使用 AsyncTask 更新 UI 线程

Android 的 AsyncTask 比 Handler 轻量一些，适用于简单的异步处理。

定义一个异步任务类 AsyncTask 的实例时，需要重写其抽象方法 doInBackground() 和另一个受保护的方法 onPostExecute()，其定义如图 8.4.2 所示。

Android Studio 移动开发教程

```
AsyncTask
    onPreExecute(): void
    doInBackground(Params...): Result
    onPostExecute(Result): void
    publishProgress(Progress...): void
    onProgressUpdate(Progress...): void
    execute(Params...): AsyncTask<Params, Progress, Result>
```

图 8.4.2 异步任务类 AsyncTask 的定义

【例 8.4.2】 使用 AsyncTask 模拟后台文件下载。

使用 AsyncTask 实现后台文件下载模拟，其效果如图 8.4.3 所示。

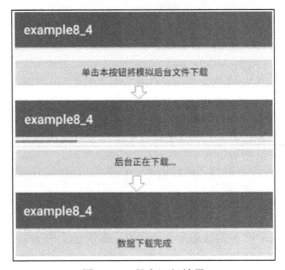

图 8.4.3 程序运行效果

程序 MainActivity.java 代码如下：

```java
/*
    使用异步任务类 AsyncTask，它是一个包含 3 个泛型参数的抽象类
    提供了多个可供重写的方法，其中 doInBackground()是必须重写的方法
*/
public class MainActivity extends AppCompatActivity {
    Button button;                          //按钮
    ProgressBar progressBar;                //进度条
    @Override
    protected void onCreate(Bundle savedInstanceState) {
        super.onCreate(savedInstanceState);
        setContentView(R.layout.activity_main);
        button = findViewById(R.id.button);
        progressBar = findViewById(R.id.progressBar);
        button.setOnClickListener(new View.OnClickListener() {
            @Override
            public void onClick(View v) {
                //执行异步任务，方法 execute()可以包含 0 或多个参数
```

```
                    new MyAsyncTask().execute();
                }
            });
        }
        class MyAsyncTask extends AsyncTask<String,Integer,String>{    //输入参数、任务进度、结果
            @Override
            protected void onPreExecute() {
                progressBar.setVisibility(View.VISIBLE);
                super.onPreExecute();
            }
            @Override
            protected String doInBackground(String... strings) {
                // 模拟耗时操作
                int step = 1;
                try {
                    while (step <= 100) {
                        //step 的值传递给 onProgressUpdate(Integer... values)中的 values 参数
                        publishProgress(step++);         //将执行 onProgressUpdate()方法
                        Thread.sleep(50);
                    }
                } catch (InterruptedException e) {
                    e.printStackTrace();
                }
                return "数据下载完成";    //返回结果传递给 onPostExecute(String s)中的 s 参数
            }
            @Override
            protected void onProgressUpdate(Integer... values) {
                progressBar.setProgress(values[0]);         //更新 ProgressBar
                super.onProgressUpdate(values);
                button.setText("后台正在下载...");
            }
            @Override
            protected void onPostExecute(String s) {       //s 参数类型与泛型第 3 个参数一致
                progressBar.setVisibility(View.GONE);      //移除 ProgressBar
                button.setText(s);                          //异步任务返回的数据更新 UI
                super.onPostExecute(s);
            }
        }
    }
}
```

8.4.4 使用 ContentProvider+AsyncTask 实现群发短信

目前，一般的手机都提供了群发短信的功能。下面介绍的案例，在选择手机联系人时由于使用复选框设计，因而操作界面更加流畅。

【例 8.4.3】 选择手机联系人后群发短信。

【程序运行】主窗口中，取消了标题栏的显示，在第一行使用水平线性布局，分别放置了用于返回的图像按钮、描述程序功能的文本框和提交选取手机联系人的图像，运行效果如图 8.4.4 所示。

图 8.4.4　程序运行效果

选择手机联系人完毕后,单击右上方的图像按钮"✓",出现确认对话框。确认后,即可进入系统提供的编辑和发送短信界面。

注意:当手机联系人较多时,出现手机联系人界面需要等待一会儿。

【设计思想】由于目前手机联系人数量较多,因此,需要使用异步任务方式获取手机联系人信息,并将获取信息和更新 UI 控件的代码写在不同的方法里。此外,列表视图要在 MainActivity 程序读取联系人后调用相关方法来呈现。

【实现步骤】

(1) 在项目的清单文件里注册读取手机联系人权限。

(2) 在程序文件夹里创建一个名为 Contact 的实体类,其代码如下:

```java
class Contact {    //自定义的实体类
    private String name;
    private String phone;
    public String getName() {
        return name;
    }
    public void setName(String name) {
        this.name = name;
    }
    public String getPhone() {
        return phone;
    }
    public void setPhone(String phone) {
        this.phone = phone;
    }
    @Override
    public String toString() {
        return "Contact{" + "name='" + name + '\'' + ", phone='" + phone + '\'' + '}';
    }
}
```

（3）在程序文件夹里创建一个包含列表视图控件，名为 ListItemView 的类，其代码如下：

```
public class ListItemView {
    ImageView iv_PersonPicture;           //图像
    TextView tv_NameAndPhone;             //姓名和手机号
    CheckBox checkBox;                    //复选框
}
```

（4）在程序文件夹里创建一个 BaseAdapter 的子类 ChooseAdapter，并重写 4 个方法。在构造方法里，使用方法 LayoutInflater.from(context)创建布局充气筒对象，以供呈现选择联系人的视图使用。此外，还要编写刷新列表视图的方法，以供 MainActivity 在获取联系人后调用，其代码如下：

```
/*
    定义数据适配器 BaseAdapter（抽象类）的子类，需要重写 4 个方法
    布局 R.layout.list_item 对应的视图对象为自定义的 ListItemView 类型
    在 BaseAdapter 的 getView()里动态呈现列表视图（响应 MainActivity 里的勾选操作）
*/
public class ChooseAdapter extends BaseAdapter {
    List<Contact> listData;    //联系人列表数据，实体类 Contact 仅包含 name 和 phone 两个字段
    LayoutInflater layoutInflater;              //布局充气筒
    HashMap<Integer, View> map = new HashMap<Integer, View>();
    // 可以改写成显示联系人照片
    static final int[] icons = {R.drawable.male, R.drawable.male};        //全局变量
    static Map<Integer, Boolean> checkResult;  //记录勾选结果，在 MainActivity 中访问
    public ChooseAdapter(Context context, List<Contact> listData) {       //构造方法
        layoutInflater = LayoutInflater.from(context);                    //创建布局充气筒对象
        this.listData = listData;
        checkResult = new HashMap<Integer, Boolean>();
        for (int i = 0; i < listData.size(); i++) {
            checkResult.put(i, false);            //默认全部未选择，作用相当于一个数组
        }
    }
    @Override
    public int getCount() {
        return listData.size();
    }
    @Override
    public Object getItem(int position) {
        return listData.get(position);
    }
    @Override
    public long getItemId(int position) {
        return position;
    }
```

```java
@Override
public View getView(int position, View convertView, ViewGroup parent) {
    ListItemView listItemView = null;
    if (convertView == null) {
        //获取 list_item 布局文件的视图
        convertView = layoutInflater.inflate(R.layout.list_item, null);
        //创建与布局 R.layout.list_item 对应的视图对象
        listItemView = new ListItemView();
        listItemView.iv_PersonPicture = convertView.findViewById(R.id.iv_PersonPicture);
        listItemView.tv_NameAndPhone = convertView.findViewById(R.id.tv_ NameAndPhone);
        listItemView.checkBox = convertView.findViewById(R.id.checkBox);
        map.put(position, convertView);           //设置控件集到 convertView
        //贴个标签
        convertView.setTag(listItemView);
    } else {
        //取出原来使用方法 setTag()贴的标签
        listItemView = (ListItemView) convertView.getTag();
    }
    Contact contactInfo = listData.get(position);   //联系人信息
    listItemView.iv_PersonPicture.setImageResource(icons[0]);
    listItemView.tv_NameAndPhone.setText(contactInfo.getName() + "   " + contactInfo.
                                                                       getPhone());
    listItemView.checkBox.setChecked(checkResult.get(position));         //Map 集合
    return convertView;
}
//供 MainActivity 调用，呈现视图
public void refresh(List<Contact> listData) {
    this.listData = listData;
    //适配器的内容改变了，通过下面的方法强制调用 getView()来刷新每个 Item 的内容
    notifyDataSetChanged();
}
}
```

（5）在 MainActivity 程序里，除了动态申请读联系人权限外，编写获取手机联系人的方法 getContacts()、呈现视图的 initView()方法，以供异步任务类 AsyncTask 的子类的相关方法调用，其代码如下：

```java
/*
    读取手机联系人的手机号码，因此，需要权限 android.permission.READ_CONTACTS
    列表适配器提供了使 MainActivity 调用的视图动态刷新方法
    调用系统的短信程序，因此，不需要短信权限
*/
public class MainActivity extends AppCompatActivity implements View.OnClickListener {
    final Uri uri=ContactsContract.Contacts.CONTENT_URI;    //Android 联系人 URI
    ListView listView;                                       //列表控件
    ChooseAdapter myAdapter;                                 //列表适配器
    List<Contact> contacts;                                  //存放联系人列表（姓名+手机号），
```

```java
List<String> receiverName = new ArrayList<String>();          //短信接收人姓名
ArrayList<String> receiverPhone = new ArrayList<String>();    //短信接收人的手机号
String receiverPhones;           //短信接收人的手机号字符串，中间以分号分隔
ImageView backBtn;               //作为返回按钮
ImageView imageButton;           //作为选择确认按钮
@Override
protected void onCreate(Bundle savedInstanceState) {
    super.onCreate(savedInstanceState);
    this.requestWindowFeature(Window.FEATURE_NO_TITLE); //取消对应用程序标题栏的
                                                                            显示
    setContentView(R.layout.activity_main);
    //权限处理
    if (ContextCompat.checkSelfPermission(this,Manifest.permission.READ_CONTACTS) !=
                                    PackageManager.PERMISSION_GRANTED){
        ActivityCompat.requestPermissions(MainActivity.this,new String[]
                                    {Manifest.permission.READ_CONTACTS},1);
    }else {
        new GetContact().execute();   //调用内部类，执行异步任务，获得联系人
        //getContacts();initView();    //阻塞式耗时，联系人个数较多时出现白屏
    }
    //控件对象监听
    backBtn = findViewById(R.id.onBack);backBtn.setOnClickListener(this);
    imageButton = findViewById(R.id.imageButton);imageButton.setOnClickListener(this);
    contacts = new ArrayList<Contact>();
}
@Override
public void onRequestPermissionsResult(int requestCode, @NonNull String[] permissions,
                                    @NonNull int[] grantResults) {
    switch (requestCode){
        case 1:
            if(grantResults.length>0){
                for(int result : grantResults){
                    if(result != PackageManager.PERMISSION_GRANTED){
                        Toast.makeText(this, "权限不足!", Toast.LENGTH_SHORT). show();
                        finish();
                        return;
                    }
                }
            }
            new GetContact().execute();   //执行异步线程
            break;
    }
}
@Override                                 //监听实现按钮功能的两个图像
public void onClick(View arg0) {
    switch (arg0.getId()) {
        case R.id.onBack:
            finish();
```

```java
                break;
            case R.id.imageButton:
                setMessageToSome();          //向选择的联系人群发短信
                break;
        }
    }
    //考虑到目前手机联系人数量增多，使用异步任务类以免阻塞 UI 主线程
    class GetContact extends AsyncTask<String, Integer, String> {
        @Override
        protected void onPreExecute() {
            super.onPreExecute();
            Toast.makeText(MainActivity.this, "联系人信息加载中，稍候...",
                                                Toast.LENGTH_SHORT).show();
        }
        @Override
        //接收单个值或数组，还可以接收离散变量
        protected String doInBackground(String... arg0) {
            getContacts();                   //获取所有联系人并存放至 contacts
            return null;
        }
        @Override
        protected void onPostExecute(String result) {
            super.onPostExecute(result);
            initView();                      //呈现视图
        }
    }
    private void getContacts() {             //获得所有联系人
        Contact tmpContact;                  //声明自定义实体类对象
        contacts.clear();
        Cursor cursor =getContentResolver().query(uri,null,null,null,
                ContactsContract.Contacts.DISPLAY_NAME + " COLLATE LOCALIZED ASC");
        if (cursor.moveToFirst()) {          //如果选择了记录
            int idColumn = cursor.getColumnIndex(ContactsContract.Contacts._ID);
            int displayNameColumn = cursor.getColumnIndex(ContactsContract.Contacts.
                                                                DISPLAY_NAME);
            do{    //循环遍历
                tmpContact = new Contact();
                String contactId = cursor.getString(idColumn);           //获得联系人 ID
                String disPlayName = cursor.getString(displayNameColumn); //获得联系人姓名
                //联系人的电话数量，没有这返回值为 0
                int phoneCount = cursor.getInt(cursor.getColumnIndex(
                                ContactsContract.Contacts.HAS_PHONE_NUMBER));
                tmpContact.setName(disPlayName);
                if (phoneCount > 0) {
                    // 获得联系人的电话号码
                    Cursor phones = getContentResolver().query(ContactsContract.
                                                                CommonDataKinds.
                            Phone.CONTENT_URI, null, ContactsContract.CommonDataKinds.
```

```java
                                    Phone.CONTACT_ID+" = " + contactId, null, null);
            if (phones.moveToFirst()) {
                String tmpPhone = phones.getString(phones.getColumnIndex(
                        ContactsContract.CommonDataKinds.Phone.NUMBER));
                //有些手机(如小米)登记手机号,多了2位空格,如 155 2764 3858
                //手机号以"1"打头
                if(tmpPhone.length() >= 11&&tmpPhone.substring(0, 1).equals("1"))
                    tmpContact.setPhone(tmpPhone);
            }
        }
        if(tmpContact.getPhone()!= null) {
            contacts.add(tmpContact);
            Log.i("myTag",tmpContact.toString());
        }else{
            Log.i("myTag","none");
        }
    } while (cursor.moveToNext());
  }
}
void initView() {                    //呈现供勾选联系人的界面
    listView = findViewById(R.id.listView);
    myAdapter = new ChooseAdapter(this, contacts);
    listView.setAdapter(myAdapter);
    listView.setItemsCanFocus(false);
    listView.setChoiceMode(ListView.CHOICE_MODE_MULTIPLE);
    //列表项选择监听,同时进行了勾选或取消勾选操作
    listView.setOnItemClickListener(new AdapterView.OnItemClickListener() {
        @Override
        public void onItemClick(AdapterView<?> parent, View view, int position, long id) {
            Toast.makeText(MainActivity.this,
                    "单击了 id 为" + position + "的联系人!", Toast.LENGTH_SHORT).
                    show();
            ListItemView listItemView = (ListItemView) view.getTag();
            boolean check = listItemView.checkBox.isChecked();
            if (check) {
                listItemView.checkBox.setChecked(false);
                check = false;
            } else {
                listItemView.checkBox.setChecked(true);
                check = true;
            }
            ChooseAdapter.checkResult.put(position, check);    //记录勾选
        }
    });
    if (contacts.size()>0) {
        listView.setClickable(true);                    //可去除,因为是默认值
        //调用适配器类定义的方法
        myAdapter.refresh(contacts);
```

```java
            }else{
                Toast.makeText(MainActivity.this, "未找到任何联系人,请检查URI,",
                                                    Toast.LENGTH_SHORT).show();
                finish();
            }
        }
    }
    public void setMessageToSome() {
        //先构建对话框内容和群发的手机号,再调用系统的短信程序,向选择的联系人发送短信
        //选择了联系人进入短信编辑界面时,可能还会按返回键重新选择
        receiverName.clear(); receiverPhone.clear();
        for (int i = 0; i < ChooseAdapter.checkResult.size(); i++) {
            if (ChooseAdapter.checkResult.get(i)) {
                receiverName.add(contacts.get(i).getName());
                receiverPhone.add(contacts.get(i).getPhone());
            }
        }
        if (receiverName.size() == 0) {                                    //未勾选联系人
            new AlertDialog.Builder(this).setTitle("没有选中任何记录").setPositiveButton("确定",
                                                                    null).show();
        } else {
            StringBuilder sb = new StringBuilder();
            sb.append("接收人:" + receiverName.get(0));
            receiverPhones = receiverPhone.get(0);
            for (int i = 1; i < receiverName.size(); i++) {
                sb.append("," + receiverName.get(i));    //构建对话框的内容、姓名之间用逗号分隔
                receiverPhones += ";" + receiverPhone.get(i);    //群发的手机号码之间用分号分隔
            }
            new AlertDialog.Builder(this)
                .setTitle("是否确定?")
                .setMessage(sb)
                .setPositiveButton("确定", new DialogInterface.OnClickListener() {
                    @Override
                    public void onClick(DialogInterface arg0, int arg1) {
                        //使用系统的短信程序
                        Intent intent=new Intent(Intent.ACTION_VIEW);
                        //短信界面中显示的接收人字符串由接收人电话字符串决定
                        intent.setData(Uri.parse("sms:"+receiverPhones+"?body=这里写群发短
                                                                    信内容"));
                        startActivity(intent);        //调用系统的短信程序
                    }
                })
                .setNegativeButton("取消", null)
                .show();
        }
    }
}
```

习 题 8

一、判断题

1．定义 ContentProvider 及使用，应属于两个不同的 Android 应用。
2．手机联系人数据包含在一张表内。
3．内容提供者程序与使用内容提供者程序操作的是同一数据源。
4．内容提供者程序的增加、删除和修改方法的返回值都是 int 类型。
5．定义 Handler 对象，必须重写方法 handleMessage()。
6．使用 Handler 对象处理消息时，发送消息必须在子线程里进行。

二、选择题

1．下列 ContentProvider 定义的_____方法，其作用相当于类的构造方法。
　　A．onCreate()　　B．insert()　　C．delete()　　D．update()
2．在项目清单文件中对内容提供者配置的<provider>标签内，不包含的属性是_____。
　　A．android:name　　　　　　B．android:authorities
　　C．android:exported　　　　D．android:label
3．检索手机联系人是否有电话，应检索手机联系人数据库中的_____表。
　　A．raw_contacts　　B．data　　C．contacts　　D．mimetypes
4．Handler 位于软件包 andoid._____内。
　　A．content　　B．app　　C．os　　D．provider
5．创建 Handler 对象时，需要实现其内部接口_____。
　　A．Callback　　B．Feedback　　C．Return　　D．Loop
6．下列不属于异步任务类 AsyncTask 提供的方法是_____。
　　A．onPreExecute()　　　　　B．doInBackground()
　　C．start()　　　　　　　　　D．onPostExecute()

三、填空题

1．在清单文件中注册自定义 ContentProvider 组件所使用的标签是_____。
2．定义内容提供者时 URI 表示访问资源的_____位置。
3．表示内容提供者时 URI 的 scheme 为_____。
4．使用内容提供者组件提供的 CRUD 方法，通过_____类型的对象调用。
5．创建 AsyncTask 的实例对象时，必须重写的方法是_____。

实 验 8

一、实验目的

（1）掌握使用 ContentProvider 建立内容提供者的方法。
（2）掌握通过 ContentResolver 访问 ContentProvider 的方法。
（3）掌握手机联系人数据库的主要表及其关系。
（4）掌握对手机联系人数据库的编程。

二、实验内容及步骤

【预备】在 Android Studio 中，新建名为 Example8 的项目后，访问本课程配套的网站 http://www.wustwzx.com/as/sy/sy08.html，复制相关代码，完成如下几个模块的设计。

1. 使用 Android 系统提供的手机联系人内容提供者

（1）在项目 Example 8 里，新建名为 example8_1 的模块。
（2）在清单文件里注册读取联系人权限 android.permission.READ_CONTACTS。
（3）MainActivity 动态申请权限，并重写回调方法 onRequestPermissionsResult()。
（4）获取内容解析器对象，以 ContactsContract.Contacts.CONTENT_URI 作为 query() 方法的参数，查询手机所有联系人信息，并存放在一个 Cursor 对象里。
（5）创建一个 SimpleAdapter 对象，添加 Cursor 数据，然后绑定到一个 ListView 控件。

2. 自定义内容提供者

（1）在项目 Example 8 里，新建名为 example8_2 的模块。
（2）在程序文件夹里，右击包名→New→Other→Content Provider，输入内容提供组件名称及应用的 URI。此时，内容提供者组件会自动在清单文件里注册。
（3）在程序文件夹里，新建名为 MyDAO 的数据库访问类，按照内容提供者组件定义的方法参数和返回值类型设计。
（4）使用类 MyDAO 的相关方法，编写内容提供者组件的相应方法。
（5）部署模块至手机，内容提供者组件的 URI 发布成功。

3. 使用自定义的内容提供者

（1）在项目 Example 8 里，新建名为 example8_2a 的模块。
（2）在布局文件里，依次添加用于输入姓名和年龄的两个 EditText 控件，表示增加、修改和删除的 3 个按钮和一个用于显示列表记录的 ListView 控件。
（3）编写程序 MainActivity.java，先获取 ContentResolver 对象，根据内容提供者的 URI，调用内容提供者的 CRUD 方法，并将查询结果显示在 ListView 控件上。
（4）运行例 8.2.1 的内容提供者程序，显示两条记录，退出再运行本内容提供者使用程序，增加一条记录后退出，再次运行内容提供者程序，可查验已经增加了一条记录。即两个程序操作相同的数据源。

4．使用 Handler 向 UI 线程传递消息

（1）在项目 Example 8 里，新建名为 example8_3 的模块。

（2）在布局文件里添加一个 Button 控件和一个 TextView 控件。

（3）在 MainActivity 里，创建一个 Handler 对象作为类成员，重写方法 handleMessage()，将接收到的信息更新到 TextView 控件。

（4）在按钮的单击事件监听器 onClick()方法里，创建一个线程并运行。在方法 run() 里，使用 Handler 对象的发送消息方法 sendMessage()。

（5）验证不使用 Hander 对象进行消息处理，直接在线程里更新 UI，运行程序时将闪退。

5．使用异步任务类 AsyncTask 更新 UI 线程

（1）在项目 Example 8 里，新建名为 example8_4 的模块。

（2）在布局文件里添加一个 ProgressBar 进度控件和一个 Button 控件。

（3）在 MainActivity 里，创建一个异步任务类 AsyncTask 的子类 MyTask，重写相关方法。

（4）在 MainActivity 的 onCreate()方法里，创建 MyAsyncTask 的实例并运行。

（5）部署本模块，做运行测试。

6．综合案例——群发短信

（1）在项目 Example 8 里，新建名为 example8_5 的模块，并在清单文件里注册读取联系人权限。

（2）在程序文件夹里创建一个名为 Contact 的实体类，包含 String 类型的两个字段 name、phone。

（3）在程序文件夹里创建一个名为 ListItemView 的类，它包含各为 iv_PersonPicture 的 ImageView 控件、名为 tv_NameAndPhone 的 TextView 控件和名为 checkBox 的控件。

（4）在程序文件夹里创建一个 BaseAdapter 的子类 ChooseAdapter extends，并重写 4 个方法。在构造方法里，使用方法 LayoutInflater.from(context)创建布局充气筒对象，以供选择联系人的视图使用。此外，还要编写刷新列表视图的方法，以供 MainActivity 在获取联系人后调用。

（5）在 MainActivity 程序里，除了动态申请读取联系人权限外，编写获取手机联系人的方法 getContacts()、呈现视图的 InitView()方法，以供异步任务类 AsyncTask 子类的相关方法调用。

三、实验小结及思考

（由学生填写，重点写上机中遇到的问题。）

第 9 章　Android 近距离通信技术

电话和短信是手机的远距离通信功能。Android 手机还有近距离的通信功能，通过手机的 WiFi 网卡、蓝牙设备和 NFC 设备等，可以实现 Android 设备之间的近距离通信。本章学习要点如下：

- 掌握手机 WiFi 的使用及 Android 的 WiFi 编程；
- 掌握手机蓝牙的使用及 Android 蓝牙编程；
- 了解 Android 手机 NFC 功能的应用。

9.1　WiFi 通信

9.1.1　WiFi 简介

WiFi（Wireless Fidelity）无线连接是一种可以将个人电脑、手持设备（如 PAD、手机）等终端以无线方式互相连接的技术。WiFi 信号也是由有线网提供的，如家里的 ADSL 和校园网等，只要接一个无线路由器，就可以把有线信号转换成高频无线电信号（WiFi 直译为无线保真）。其中，WiFi 地址是热点无线路由器发出信号的地址，WiFi MAC 是终端的物理地址，在局域网中辨别终端的标识，它们均为 6 个字节。

WiFi 上网方式，因不使用移动网络而节省了流量费，因不受布线条件的限制非常适合移动办公用户的需要。此外，WiFi 发射信号功率低于 100mW，比手机发射功率低。所以，WiFi 上网也是相对安全和健康的。

WiFi 是一种局域网协议，它是为改善基于 IEEE 802.11 标准的无线网络产品之间的互通性而出现的，其工作频段为 2.4 GHz，传输速率可以达到 54Mb/s。

出于安全考虑，每个无线路由器里都可以设置其使用 WiFi 的密码，它是用于登录无线网络的依据。

注意：在手机检测到的 WiFi 信号中，带锁的图标则表示设置了使用密码。

9.1.2　Android 对 WiFi 的支持

在包 android.net 内提供了 3 个 WiFi 类：管理器类 WifiManager、扫描结果类 ScanResult 和信息类 WifiInfo。

WifiManager 主要提供了获取 WiFi 状态、打开或关闭 WiFi、扫描 WiFi 并获取结果

第9章 Android近距离通信技术

和获得当前WiFi连接等方法，如图9.1.1所示。

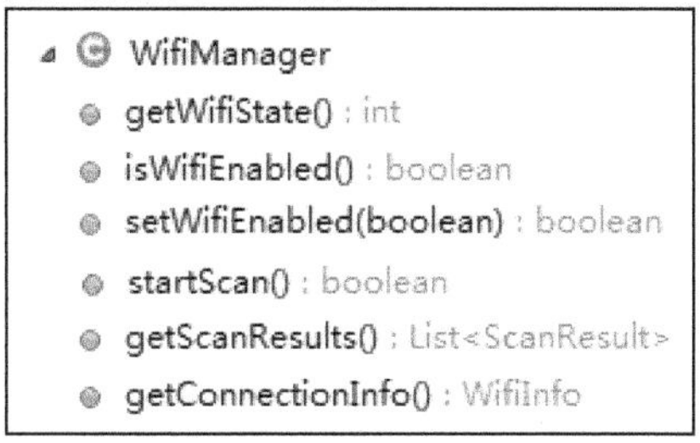

图9.1.1 WifiManager的常用方法

在调用WifiManager类的getScanResults()方法前，必须先使用WifiManager类的startScan()方法，扫描结果是一个列表类型List<ScanResult>。

ScanResult类包含了信源的一些基本信息，主要有WiFi名称（对应于SSID字段）、WiFi地址（对应于BSSID字段，表示无线路由器的MAC地址）、WiFi频率（对应于frequency字段）和WiFi信号强度（对应于level字段）等，如图9.1.2所示。

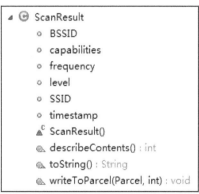

图9.1.2 ScanResult类的定义

通过WifiManager类的getConnectionInfo()方法实时获取当前连接WiFi的有关信息，其结果是一个WifiInfo类型。

WifiInfo类提供了获取当前WiFi名称方法getSSID()、WiFi路由器MAC地址（接入点地址）方法getBSSID()、WiFi信号强度方法getRssi()、手机WiFi网卡的MAC地址（本机地址）方法getMacAddress()等，如图9.1.3所示。

图9.1.3 WifiInfo类的常用方法

注意：经比较可以看出，类 WifiInfo 中的某些方法与类 ScanResult 的某些属性相对应。

9.1.3 WiFi 应用实例

下面介绍一个例子，以掌握对 WiFi 相关类的使用。

【例 9.1.1】 WiFi 的基本操作。

【程序运行】程序运行时，主界面上出现 4 个功能按钮。单击"检查 WiFi"按钮，会出现一个显示 WiFi 状态的 Toast 消息。如果 WiFi 没有开启，单击"打开 WiFi"按钮，然后单击"扫描 WiFi"按钮，其扫描结果在一个 TextView 控件内显示，程序运行效果如图 9.1.4 所示。

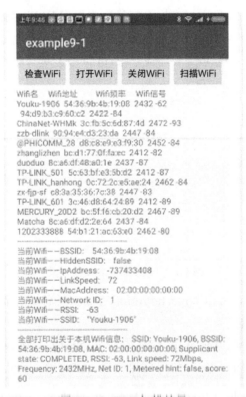

图 9.1.4　WiFi 扫描结果

主要设计步骤如下。

（1）先新建名为 Example9 的项目，再新建名为 examle9_1 的模块。

（2）在模块的清单文件里，注册 WiFi 相关权限，其代码如下：

```
<!--下面 2 个是普通权限，只需要在清单文件里注册，不需要在程序里动态申请-->
<uses-permission android:name="android.permission.CHANGE_WIFI_STATE" />
<uses-permission android:name="android.permission.ACCESS_WIFI_STATE" />
<!--搜索 WiFi 时需要定位权限，它是危险权限，还需要在程序里动态申请-->
<uses-permission android:name="android.permission.ACCESS_COARSE_LOCATION"/>
```

（3）在模块的布局文件中，添加 4 个操作 WiFi 的 Button 按钮、1 个产生卷动效果的 ScrollView 容器控件并内嵌入 1 个用于显示 WiFi 信息的 TextView 控件。

注意：当扫描到的 WiFi 信源较多使一屏显示不完全时，可在布局文件中使用实现卷轴效果的容器控件 ScrollView。其实，也可以设置文本框的手势动作。

（4）编写 Activity 组件 MainActivity，其代码如下：

```java
/*
    本程序演示了 WiFi 操作
    WiFi 权限是普通权限，对应于 WiFi 的打开、关闭和状态读取。
    WiFi 扫描需要定位权限，因此，除了在清单文件里注册外，还需要在程序里动态注册
*/
public class MainActivity extends AppCompatActivity implements OnClickListener{
    WifiManager wifiManager;
    WifiInfo wifiInfo;
    ScanResult scanResult;
    List<ScanResult> WifiList;
    Button btn_check, btn_open, btn_close, btn_search;
    TextView textView;
    public void onCreate(Bundle savedInstanceState) {
        super.onCreate(savedInstanceState);
        //设定竖屏，不会切换成横屏
        setRequestedOrientation(ActivityInfo.SCREEN_ORIENTATION_PORTRAIT);
        setContentView(R.layout.activity_main);
        if (ContextCompat.checkSelfPermission(this,Manifest.permission.ACCESS_COARSE_
                                                                        LOCATION) !=
                                        PackageManager.PERMISSION_GRANTED) {
            ActivityCompat.requestPermissions(this,new String[]{
                                Manifest.permission.ACCESS_COARSE_LOCATION}, 1);
        }
        fun();
    }
    private void fun() {
        btn_check = findViewById(R.id.btn_check);btn_check.setOnClickListener(this);
        btn_open = findViewById(R.id.btn_open);btn_open.setOnClickListener(this);
        btn_close = findViewById(R.id.btn_close);btn_close.setOnClickListener(this);
        btn_search = findViewById(R.id.btn_search);btn_search.setOnClickListener(this);
        textView = findViewById(R.id.textView);
        //创建 WiFi 管理器对象
        wifiManager = (WifiManager) getApplicationContext().getSystemService(Context.WIFI_
                                                                        SERVICE);
    }
    @Override
    public void onClick(View v) {    //内部接口 View.OnClickListener 要实现的方法
        int ItemId = v.getId();//获取控件的 Id 值
        switch (ItemId) {
```

```java
case R.id.btn_check:
    textView.setVisibility(View.INVISIBLE); //隐藏先前扫描的结果
    Toast.makeText(MainActivity.this, "当前 WiFi 状态为: " + getWifiState(),
                        Toast.LENGTH_LONG).show();
    break;
case R.id.btn_open:   //打开 WiFi, 存在延时
    textView.setVisibility(View.INVISIBLE);
    wifiManager.setWifiEnabled(true);   //本方法在 Android 10 设备上失效
    long start=System.currentTimeMillis();   //记录开始时间
    while (true){
        if(wifiManager.getWifiState()==3){
            break;   //中止循环
        }
    }
    long stop=System.currentTimeMillis();   //记录结束时间
    Toast.makeText(MainActivity.this, "当前 WiFi 状态为: "+getWifiState()+
        "\n 打开 WiFi 花费"+(stop-start)+"毫秒", Toast.LENGTH_LONG).show();
    break;
case R.id.btn_close:  //关闭 WiFi, 存在延时
    textView.setVisibility(View.INVISIBLE);
    wifiManager.setWifiEnabled(false);   //本方法在 Android 10 设备上失效
    long start2=System.currentTimeMillis();   //记录开始时间
    while (true){
        if(wifiManager.getWifiState()==1){
            break;   //中止循环
        }
    }
    long stop2=System.currentTimeMillis();   //记录结束时间
    Toast.makeText(MainActivity.this, "当前 WiFi 状态为: "+getWifiState()+
        "\n 关闭 WiFi 花费"+(stop2-start2)+"毫秒", Toast.LENGTH_LONG).show();
    break;
case R.id.btn_search:
    wifiManager.startScan();   //开始扫描
    WifiList = wifiManager.getScanResults();   //获取扫描结果
    wifiInfo = wifiManager.getConnectionInfo();   //当前连接的 WiFi 信息
    StringBuffer stringBuffer = new StringBuffer();
    stringBuffer
            .append("Wifi 名").append("              ")
            .append("Wifi 地址").append("                  ")
            .append("Wifi 频率").append("      ")
            .append("Wifi 信号\n");
    if (WifiList != null) {
        for (int i = 0; i < WifiList.size(); i++) {
            scanResult = WifiList.get(i);
            stringBuffer
                    .append(scanResult.SSID).append("   ")
                    .append(scanResult.BSSID).append("   ")
                    .append(scanResult.frequency).append(" ")
                    .append(scanResult.level).append("\n");
        }
        stringBuffer
```

```
                        .append("----------------------------------------------\n")
                        .append("本机使用 WiFi 的相关指标:\n")
                        .append("BSSID: ").append(wifiInfo.getBSSID()).append("\n")
                        .append("Hidden SSID: ").append(wifiInfo.getHiddenSSID()).
                                                                         append("\n")
                        .append("IP Address: ").append(wifiInfo.getIpAddress()).
                                                                         append("\n")
                        .append("Link Speed: ").append(wifiInfo.getLinkSpeed()).
                                                                         append("\n")
                        .append("MAC Address: ").append(wifiInfo.getMacAddress()).
                                                                         append("\n")
                        .append("Network ID: ").append(wifiInfo.getNetworkId()).
                                                                         append("\n")
                        .append("RSSI: ").append(wifiInfo.getRssi()).append("\n")
                        .append("SSID: ").append(wifiInfo.getSSID()).append("\n");
                stringBuffer
                        .append("----------------------------------------------\n")
                        .append("本机使用 WiFi 的信息:\n")
                        .append(wifiInfo.toString());
                textView.setVisibility(View.VISIBLE);
                textView.setText(stringBuffer.toString());
            }
        }
    }
    public String getWiFiState(){       //自定义获取 WiFi 状态方法
        String temp=null;
        switch (wifiManager.getWifiState()) {   //与自定义方法同名
            case 0:
                temp="WiFi 正在关闭...";
                break;
            case 1:
                temp="WiFi 已经关闭";
                break;
            case 2:
                temp="WiFi 正在打开...";
                break;
            case 3:
                temp="WiFi 已经打开";
                break;
            default:
                break;
        }
        return temp;
    }
    @Override
    public void onRequestPermissionsResult(int requestCode, @NonNull String[] permissions,
```

```
                                            @NonNull int[] grantResults) {
    super.onRequestPermissionsResult(requestCode, permissions, grantResults);
    if(grantResults.length>0){
        if(grantResults[0]!=PackageManager.PERMISSION_GRANTED){
            Toast.makeText(this, "未授权，WiFi 搜索功能将不可用！",Toast.LENGTH_
                                                                SHORT).show();
        }else{
            fun();
        }
    }
}
```

9.2 蓝牙通信Bluetooth

9.2.1 Bluetooth 简介

蓝牙（Bluetooth）是一种支持设备短距离（一般 10 m 内）通信的无线电技术，能在包括移动电话、PDA、无线耳机、笔记本电脑、相关外设等众多设备之间进行无线信息交换。

蓝牙采用分散式网络结构、快跳频和短包技术，支持点对点及一点对多点通信，工作在全球通用的 2.4GHz ISM（工业、科学、医学）频段，其数据速率为 1 Mb/s。蓝牙采用时分双工传输方案实现全双工传输。

注意：

（1）蓝牙通信可以是双向的，这不同于 WiFi 通信。另外，蓝牙通信与 WiFi 使用的频段也不同。

（2）蓝牙鼠标、车载电话等，都是蓝牙技术的应用。

9.2.2 Android 对 Bluetooth 的支持

蓝牙适配器就是各种数码产品能适用蓝牙设备的接口转换器，它采用了全球通用的短距离无线连接技术。利用蓝牙技术，能够有效地简化移动通信终端设备之间的通信，也能够成功地简化设备与 Internet 之间的通信，从而使数据传输变得更加迅速高效，为无线通信拓宽了道路。

Android 包 android.bluetooth 提供了蓝牙的相关类，其主要类如图 9.2.1 所示。

蓝牙适配器类 BluetoothAdapter 代表本地的蓝牙适配器设备，可以使用户执行基本的蓝牙任务。如初始化设备的搜索、查询可匹配的设备集、使用一个已知的 MAC 地址来初始化一

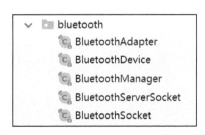

图 9.2.1 android.bluetooth 包里的主要类

个 BluetoothDevice 类、创建一个 BluetoothServerSocket 类以监听其他设备对本机的连接请求等。

使用静态方法 BluetoothAdapter.getDefaultAdapter()，能够获取代表本地蓝牙适配器的 BluetoothAdapter 对象。在拥有本地适配器以后，用户可以获得一系列的 BluetoothDevice 对象，这些对象代表所有拥有 getBondedDevices()方法且已经匹配的设备；用 startDiscovery()方法开始设备的搜寻；或者创建一个 BluetoothServerSocket 类，通过 listenUsingRfcommWithServiceRecord(String, UUID)方法监听新来的连接请求。BluetoothAdapter 类的定义，如图 9.2.2 所示。

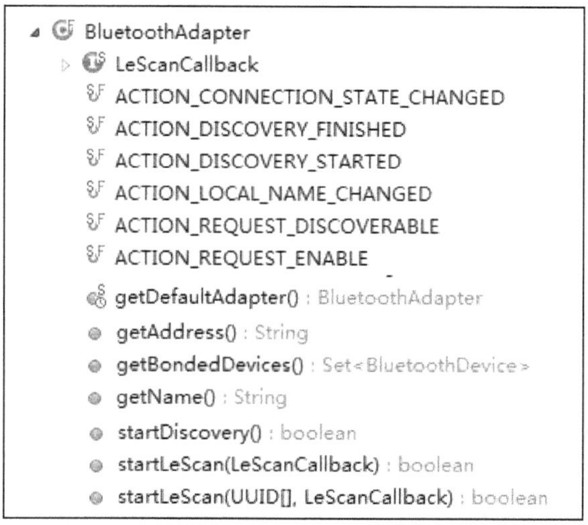

```
▲ ⓒ BluetoothAdapter
    ▷ ⓒ LeScanCallback
      ⓕ ACTION_CONNECTION_STATE_CHANGED
      ⓕ ACTION_DISCOVERY_FINISHED
      ⓕ ACTION_DISCOVERY_STARTED
      ⓕ ACTION_LOCAL_NAME_CHANGED
      ⓕ ACTION_REQUEST_DISCOVERABLE
      ⓕ ACTION_REQUEST_ENABLE
      ⓜ getDefaultAdapter() : BluetoothAdapter
      ⓜ getAddress() : String
      ⓜ getBondedDevices() : Set<BluetoothDevice>
      ⓜ getName() : String
      ⓜ startDiscovery() : boolean
      ⓜ startLeScan(LeScanCallback) : boolean
      ⓜ startLeScan(UUID[], LeScanCallback) : boolean
```

图 9.2.2　BluetoothAdapter 类的定义

注意：智能手机一般都配有蓝牙适配器，但大部分台式计算机没有蓝牙适配器。要想实现计算机与手机的蓝牙通信，需要在计算机上添加蓝牙适配器。目前，USB 型的蓝牙适配器为计算机增加了蓝牙功能。

蓝牙管理器类用来管理远程蓝牙设备，主要包括显示所有蓝牙配对设备的列表、配对新设备的扫描、设置本机的蓝牙设备等。BluetoothManager 的定义如图 9.2.3 所示。

```
▲ ⓒ BluetoothManager
    ⓒ BluetoothManager()
    ⓜ getAdapter() : BluetoothAdapter
    ⓜ getConnectedDevices(int) : List<BluetoothDevice>
    ⓜ getConnectionState(BluetoothDevice, int) : int
    ⓜ getDevicesMatchingConnectionStates(int, int[]) : List<BluetoothDevice>
    ⓜ openGattServer(Context, BluetoothGattServerCallback) : BluetoothGattServer
```

图 9.2.3　BluetoothManager 的定义

蓝牙设备类 BluetoothDevice 的定义如图 9.2.4 所示。

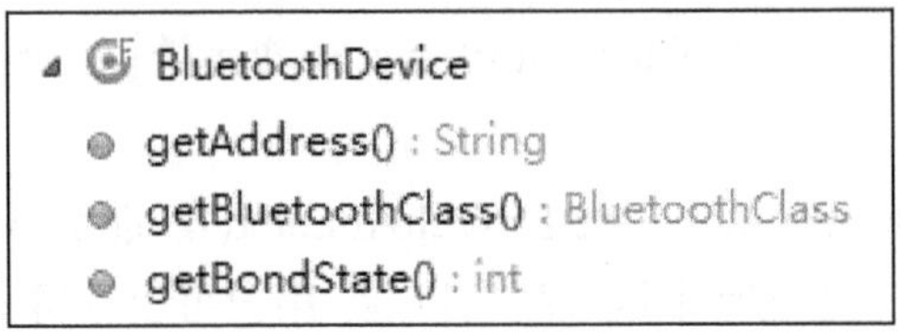

```
▲ G BluetoothDevice
    ⊚ getAddress() : String
    ⊚ getBluetoothClass() : BluetoothClass
    ⊚ getBondState() : int
```

图 9.2.4　BluetoothDevice 的定义

用于蓝牙 Socket 通信的两类分别为 BluetoothSocket（客户端，即发出请求的那一端）和 BluetoothServerSocket（服务器端，即被请求连接的那一端）。

BluetoothSocket 类的定义，如图 9.2.5 所示。

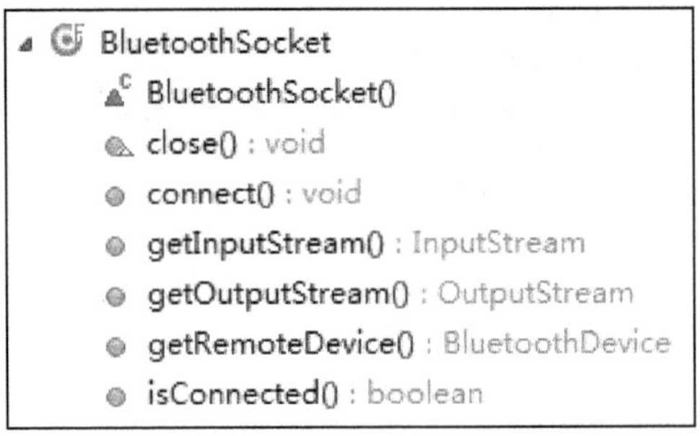

```
▲ G BluetoothSocket
    ▲c BluetoothSocket()
    ⊚ close() : void
    ⊚ connect() : void
    ⊚ getInputStream() : InputStream
    ⊚ getOutputStream() : OutputStream
    ⊚ getRemoteDevice() : BluetoothDevice
    ⊚ isConnected() : boolean
```

图 9.2.5　BluetoothSocket 类的定义

BluetoothServerSocket 类的定义，如图 9.2.6 所示。

```
▲ G BluetoothServerSocket
    ▲c BluetoothServerSocket()
    ⊚ accept() : BluetoothSocket
    ⊚ accept(int) : BluetoothSocket
    ⊚ close() : void
```

图 9.2.6　BluetoothServerSocket 类的定义

注意：

（1）手机蓝牙会话的服务程序，使用了 BluetoothSocket 建立蓝牙连接，这与 Java 的 Socket 通信是类似的。

（2）蓝牙 Socket 通信是比较底层的网络编程，是跨平台的编程方式。

在应用程序中，请求或建立蓝牙连接并传递数据，需要使用 BLUETOOTH 权限；初始化设备发现功能或更改蓝牙设置，则需要 BLUETOOTH_ADMIN 权限和定位权限 ACCESS_COARSE_LOCATION。在清单文件中，需要配置与蓝牙相关权限的代码如下：

```xml
<!--下面 2 个是普通权限，只需要在清单文件里注册，不需要在程序里动态申请-->
<uses-permission android:name="android.permission.BLUETOOTH_ADMIN" />
<uses-permission android:name="android.permission.BLUETOOTH" />
<!--Android 6.0 及以上版本的蓝牙设备搜索，需要定位权限，它是危险权限-->
<uses-permission android:name="android.permission.ACCESS_COARSE_LOCATION"/>
```

第9章　Android近距离通信技术

当蓝牙开关未打开时，使用请求的代码如下：

```
if (!bluetoothAdapter.isEnabled()) {
    //创建开启蓝牙请求的意图对象
    Intent intent = new Intent(BluetoothAdapter.ACTION_REQUEST_ENABLE);
    //请求开启蓝牙
    startActivityForResult(intent, REQUEST_ENABLE_BT);
}
```

执行上述代码的效果，如图9.2.7所示。

图9.2.7　请求打开蓝牙功能

9.2.3　蓝牙聊天实例

利用手机的蓝牙功能，可以实现在两个蓝牙手机之间的聊天功能。

注意：手机蓝牙通信并不需要移动数据网络或WiFi网络的支持。

【例9.2.1】　使用手机蓝牙实现聊天功能。

将应用程序分别部署到两个手机并运行，窗口标题栏显示"无连接"。单击手机菜单键后，选择"我的好友"菜单项，进入另一个选择已经配对好友的界面。选择一个好友后，将返回至主界面，并在标题栏右边显示已经连接的手机。此时，即可与连接的手机使用蓝牙聊天，效果如图9.2.8所示。

图9.2.8　两个手机蓝牙连接后的聊天界面

主要设计步骤如下：

（1）在项目Example9里，新建名为exaple9_2的模块。在模块清单文件里，添加注册本应用所需要的相关权限。

（2）在文件res/values/strings.xml里，添加程序运行过程中的状态描述文本及配色代码等，其代码如下：

```xml
<?xml version="1.0" encoding="utf-8"?>
<resources>
    <string name="app_name">蓝牙 Demo</string>
    <string name="send">发送</string>
    <string name="not_connected">你没有链接一个设备</string>
    <string name="bt_not_enabled_leaving">蓝牙不可用，离开聊天室</string>
    <string name="title_connecting">链接中...</string>
    <string name="title_connected_to">连接到：</string>
    <string name="title_not_connected">无链接</string>
    <string name="scanning">搜索好友中...</string>
    <string name="select_device">选择一个好友链接</string>
    <string name="none_paired">没有配对好友</string>
    <string name="none_found">附近没有发现好友</string>
    <string name="title_paired_devices">配对好友</string>
    <string name="title_other_devices">其他可连接好友</string>
    <string name="button_scan">搜索好友</string>
    <string name="connect">我的好友</string>
    <string name="discoverable">设置在线</string>
    <string name="back">退出</string>
    <string name="startVideo">开始聊天</string>
    <string name="stopVideo">结束聊天</string>
    <drawable name="bg">#ADD8E6</drawable>
    <drawable name="bg01">#87CEFA</drawable>
</resources>
```

（3）在默认的主布局文件 activity_main.xml 里，添加 1 个 Toolbar 控件，其内包含 2 个水平的 TextView 控件；在 Toolbar 控件的下方添加 1 个 ListView 控件，用于显示聊天内容；最后在 ListView 控件的下方添加水平放置的 1 个 EditText 控件和 1 个 Button 控件。使用垂直线性布局并嵌套水平线性布局实现的代码如下：

```xml
<?xml version="1.0" encoding="utf-8"?>
<LinearLayout xmlns:android="http://schemas.android.com/apk/res/android"
    android:orientation="vertical"
    android:layout_width="match_parent"
    android:layout_height="match_parent"
    android:background="@color/bg01">
    <!--新版 Android 支持的 Toolbar，对标题栏布局-->
    <android.support.v7.widget.Toolbar
        android:id="@+id/toolbar"
        android:layout_width="match_parent"
        android:layout_height="wrap_content">
        <LinearLayout
            android:layout_width="match_parent"
            android:layout_height="match_parent"
            android:orientation="horizontal">
```

```xml
        <TextView
            android:id="@+id/title_left_text"
            style="?android:attr/windowTitleStyle"
            android:layout_width="0dp"
            android:layout_height="match_parent"
            android:layout_alignParentLeft="true"
            android:layout_weight="1"
            android:gravity="left"
            android:ellipsize="end"
            android:singleLine="true" />
        <TextView
            android:id="@+id/title_right_text"
            android:layout_width="0dp"
            android:layout_height="match_parent"
            android:layout_alignParentRight="true"
            android:layout_weight="1"
            android:ellipsize="end"
            android:gravity="right"
            android:singleLine="true"
            android:textColor="#fff" />
    </LinearLayout>
</android.support.v7.widget.Toolbar>
<ListView android:id="@+id/in"
    android:layout_width="match_parent"
    android:layout_height="match_parent"
    android:stackFromBottom="true"
    android:transcriptMode="alwaysScroll"
    android:layout_weight="1" />
<LinearLayout
    android:orientation="horizontal"
    android:layout_width="match_parent"
    android:layout_height="wrap_content" >
    <EditText android:id="@+id/edit_text_out"
        android:layout_width="wrap_content"
        android:layout_height="wrap_content"
        android:layout_weight="1"
        android:layout_gravity="bottom" />
    <Button android:id="@+id/button_send"
        android:layout_width="wrap_content"
        android:layout_height="wrap_content"
        android:text="@string/send"/>
</LinearLayout>
</LinearLayout>
```

（4）编写用于蓝牙会话的服务组件 ChatService，其文件代码如下：

```java
/*
    本程序 ChatService 是蓝牙会话的服务程序
    定义 3 个线程类作为内部类：AcceptThread（接受新连接）、ConnectThread（发出连接）
        和 ConnectedThread （已连接）
*/
public class ChatService {
    private static final String NAME = "BluetoothChat";
    // UUID：通用唯一识别码，是一个 128 位长的数字，一般用十六进制表示
    //算法的核心思想是结合机器的网卡、当地时间、一个随机数来生成
    private static final UUID MY_UUID = UUID.fromString("fa87c0d0-afac-11de-8a39-
                                                        0800200c9a66");

    private final BluetoothAdapter mAdapter;
    private final Handler mHandler;
    private AcceptThread mAcceptThread;
    private ConnectThread mConnectThread;
    private ConnectedThread mConnectedThread;
    private int mState;
    public static final int STATE_NONE = 0;
    public static final int STATE_LISTEN = 1;
    public static final int STATE_CONNECTING = 2;
public static final int STATE_CONNECTED = 3;
//构造方法，接收 UI 主线程传递的对象
    public ChatService(Context context, Handler handler) {
        mAdapter = BluetoothAdapter.getDefaultAdapter();
        mState = STATE_NONE;
        mHandler = handler;
    }
    private synchronized void setState(int state) {
        mState = state;
        mHandler.obtainMessage(BluetoothChat.MESSAGE_STATE_CHANGE, state, -1).
                                                                    sendToTarget();
    }
    public synchronized int getState() {
        return mState;
    }
    public synchronized void start() {
        if (mConnectThread != null) {
            mConnectThread.cancel();
            mConnectThread = null;
        }
        if (mConnectedThread != null) {
            mConnectedThread.cancel();
            mConnectedThread = null;
        }
        if (mAcceptThread == null) {
            mAcceptThread = new AcceptThread();
```

```
            mAcceptThread.start();
        }
        setState(STATE_LISTEN);
    }
    //取消 CONNECTING 和 CONNECTED 状态下的相关线程，然后运行新的
                                                         mConnectThread 线程
    public synchronized void connect(BluetoothDevice device) {
        if (mState == STATE_CONNECTING) {
            if (mConnectThread != null) {
                mConnectThread.cancel();
                mConnectThread = null;
            }
        }
        if (mConnectedThread != null) {
            mConnectedThread.cancel();
            mConnectedThread = null;
        }
        mConnectThread = new ConnectThread(device);
        mConnectThread.start();
        setState(STATE_CONNECTING);
    }
    /*
        开启一个 ConnectedThread 来管理对应的当前连接。应先取消任意现存的
                                                         mConnectThread 、
        mConnectedThread 、 mAcceptThread 线程，然后开启新 mConnectedThread ，
        传入当前刚刚接受的 Socket 连接，最后通过 Handler 来通知 UI 连接
    */
    public synchronized void connected(BluetoothSocket socket, BluetoothDevice device) {
        if (mConnectThread != null) {
            mConnectThread.cancel();
            mConnectThread = null;
        }
        if (mConnectedThread != null) {
            mConnectedThread.cancel();
            mConnectedThread = null;
        }
        if (mAcceptThread != null) {
            mAcceptThread.cancel();
            mAcceptThread = null;
        }
        mConnectedThread = new ConnectedThread(socket);
        mConnectedThread.start();
        Message msg = mHandler.obtainMessage(BluetoothChat.MESSAGE_DEVICE_NAME);
        Bundle bundle = new Bundle();
        bundle.putString(BluetoothChat.DEVICE_NAME, device.getName());
        msg.setData(bundle);
```

```java
        mHandler.sendMessage(msg);
        setState(STATE_CONNECTED);
}
// 停止所有相关线程，设当前状态为 NONE
public synchronized void stop() {
    if (mConnectThread != null) {
        mConnectThread.cancel();
        mConnectThread = null;
    }
    if (mConnectedThread != null) {
        mConnectedThread.cancel();
        mConnectedThread = null;
    }
    if (mAcceptThread != null) {
        mAcceptThread.cancel();
        mAcceptThread = null;
    }
    setState(STATE_NONE);
}
//在 STATE_CONNECTED 状态下，调用 mConnectedThread 里的 write()，写入 byte
public void write(byte[] out) {
    ConnectedThread r;
    synchronized (this) {
        if (mState != STATE_CONNECTED)
            return;
        r = mConnectedThread;
    }
    r.write(out);
}
//连接失败时的处理，通知 UI 主线程，并设为 STATE_LISTEN 状态
private void connectionFailed() {
    setState(STATE_LISTEN);
    Message msg = mHandler.obtainMessage(BluetoothChat.MESSAGE_TOAST);
    Bundle bundle = new Bundle();
    bundle.putString(BluetoothChat.TOAST, "链接不到设备");
    msg.setData(bundle);
    //向主线程发送消息
    mHandler.sendMessage(msg);
}
// 当连接失去的时候，设为 STATE_LISTEN 状态并通知 UI
private void connectionLost() {
    setState(STATE_LISTEN);
    Message msg = mHandler.obtainMessage(BluetoothChat.MESSAGE_TOAST);
    Bundle bundle = new Bundle();
    bundle.putString(BluetoothChat.TOAST, "设备链接中断");
    msg.setData(bundle);
```

```java
            mHandler.sendMessage(msg);
        }
// 创建监听线程，准备接受新连接。使用阻塞方式，调用 BluetoothServerSocket.accept()
private class AcceptThread extends Thread {
    private final BluetoothServerSocket mmServerSocket;
    public AcceptThread() {
        BluetoothServerSocket tmp = null;
        try {
            tmp = mAdapter.listenUsingRfcommWithServiceRecord(NAME, MY_UUID);
        } catch (IOException e) {
        }
        mmServerSocket = tmp;
    }
    @Override
    public void run() {
        setName("AcceptThread");
        BluetoothSocket socket = null;
        while (mState != STATE_CONNECTED) {
            try {
                socket = mmServerSocket.accept();
            } catch (IOException e) {
                break;
            }
            if (socket != null) {
                synchronized (ChatService.this) {
                    switch (mState) {
                        case STATE_LISTEN:
                        case STATE_CONNECTING:
                            connected(socket, socket.getRemoteDevice());
                            break;
                        case STATE_NONE:
                        case STATE_CONNECTED:
                            try {
                                socket.close();
                            } catch (IOException e) {
                                e.printStackTrace();
                            }
                            break;
                    }
                }
            }
        }
    }
    public void cancel() {
        try {
            mmServerSocket.close();
```

```
            } catch (IOException e) {
                e.printStackTrace();
            }
        }
    }
    /*
    连接线程,专门用来对外发出连接对方蓝牙的请求和处理流程。
    构造函数里通过 BluetoothDevice 类方法 createRfcommSocketToServiceRecord(),
    从待连接的 BluetoothDevice 对象产生 BluetoothSocket. 然后在 run()方法中进行连接,
    成功后调用 BluetoothChatSevice 的 connected() 方法。
    定义 cancel() 在关闭线程时能够关闭相关 BluetoothSocket。
    */
    private class ConnectThread extends Thread {
        private final BluetoothSocket mmSocket;
        private final BluetoothDevice mmDevice;
        public ConnectThread(BluetoothDevice device) {
            mmDevice = device;
            BluetoothSocket tmp = null;
            try {
                tmp = device.createRfcommSocketToServiceRecord(MY_UUID);
            } catch (IOException e) {
                e.printStackTrace();
            }
            mmSocket = tmp;
        }
        @Override
        public void run() {
            setName("ConnectThread");
            mAdapter.cancelDiscovery();
            try {
                mmSocket.connect();
            } catch (IOException e) {
                connectionFailed();
                try {
                    mmSocket.close();
                } catch (IOException e2) {
                    e.printStackTrace();
                }
                ChatService.this.start();
                return;
            }
            synchronized (ChatService.this) {
                mConnectThread = null;
            }
            connected(mmSocket, mmDevice);
        }
```

```java
        public void cancel() {
            try {
                mmSocket.close();
            } catch (IOException e) {
                e.printStackTrace();
            }
        }
    }
    /*
        双方蓝牙连接后一直运行的线程。在构造函数中设置输入、输出流。
        run()方法中使用阻塞模式的 InputStream.read()循环读取输入流,
        然后发送到 UI 线程中更新聊天消息。
        本线程也提供了 write()将聊天消息写入输出流传输至对方,传输成功后回写入 UI 线程。
        最后使用 cancel()关闭连接的 Socket
    */
    private class ConnectedThread extends Thread {
        private final BluetoothSocket mmSocket;
        private final InputStream mmInStream;
        private final OutputStream mmOutStream;
        public ConnectedThread(BluetoothSocket socket) {
            mmSocket = socket;
            InputStream tmpIn = null;
            OutputStream tmpOut = null;
            try {
                tmpIn = socket.getInputStream();
                tmpOut = socket.getOutputStream();
            } catch (IOException e) {
                e.printStackTrace();
            }
            mmInStream = tmpIn;
            mmOutStream = tmpOut;
        }
        @Override
        public void run() {
            byte[] buffer = new byte[1024];
            int bytes;
            while (true) {
                try {
                    bytes = mmInStream.read(buffer);
                    mHandler.obtainMessage(BluetoothChat.MESSAGE_READ, bytes, -1,
                                            buffer).sendToTarget();
                } catch (IOException e) {
                    connectionLost();
                    break;
                }
```

```
            }
        }
        public void write(byte[] buffer) {
            try {
                mmOutStream.write(buffer);
                mHandler.obtainMessage(BluetoothChat.MESSAGE_WRITE, 1, 1, buffer).
                                                                     sendToTarget();
            } catch (IOException e) {
                e.printStackTrace();
            }
        }
        public void cancel() {
            try {
                mmSocket.close();
            } catch (IOException e) {
                e.printStackTrace();
            }
        }
    }
}
```

（5）分别建立供主 Activity 使用的菜单文件 res/menu/option_menu.xml、选择好友（已经配对过的蓝牙设备）的界面布局文件 device_list.xml。

菜单文件 option_menu.xml 的代码如下：

```xml
<?xml version="1.0" encoding="utf-8"?>
<menu xmlns:android="http://schemas.android.com/apk/res/android">
    <item android:id="@+id/scan"
          android:icon="@android:drawable/ic_menu_myplaces"
          android:title="@string/connect" />
    <item android:id="@+id/discoverable"
          android:icon="@android:drawable/ic_menu_view"
          android:title="@string/discoverable" />
    <item android:id="@+id/back"
          android:icon="@android:drawable/ic_menu_close_clear_cancel"
          android:title="@string/back" />
</menu>
```

选择好友界面的布局文件 device_list.xml 的代码如下：

```xml
<?xml version="1.0" encoding="utf-8"?>
<LinearLayout xmlns:android="http://schemas.android.com/apk/res/android"
    android:orientation="vertical"
    android:layout_width="match_parent"
    android:layout_height="match_parent">
    <TextView android:id="@+id/title_paired_devices"
        android:layout_width="match_parent"
```

```xml
        android:layout_height="wrap_content"
        android:text="@string/title_paired_devices"
        android:visibility="gone"
        android:background="#666"
        android:textColor="#fff"
        android:paddingLeft="5dp" />
    <ListView android:id="@+id/paired_devices"
        android:layout_width="match_parent"
        android:layout_height="wrap_content"
        android:layout_weight="1" />
    <TextView android:id="@+id/title_new_devices"
        android:layout_width="match_parent"
        android:layout_height="wrap_content"
        android:text="@string/title_other_devices"
        android:visibility="gone"
        android:background="#666"
        android:textColor="#fff"
        android:paddingLeft="5dp" />
    <!--android:visibility="gone"表示不占空间的隐藏,invisible 是占空间-->
    <ListView android:id="@+id/new_devices"
        android:layout_width="match_parent"
        android:layout_height="wrap_content"
        android:layout_weight="2" />
    <Button android:id="@+id/button_scan"
        android:layout_width="match_parent"
        android:layout_height="wrap_content"
        android:text="@string/button_scan" />
</LinearLayout>
```

(6) 新建 Activity 组件 DeviceList,实现选取与之会话的蓝牙设备,其文件代码如下:

```java
/*
    本程序供菜单项主界面的选项菜单"我的友好"调用,用于:
    (1) 显示已配对的好友列表
    (2) 搜索可配对的好友进行配对
    (3) 新选择并配对的蓝牙设备将刷新好友列表
    注意:发现新的蓝牙设备并请求配对时,需要对应接受
    关键技术:动态注册一个广播接收者,处理蓝牙设备扫描的结果
*/
public class DeviceList extends AppCompatActivity{
    private BluetoothAdapter mBtAdapter;
    private ArrayAdapter<String> mPairedDevicesArrayAdapter;
    private ArrayAdapter<String> mNewDevicesArrayAdapter;
    public static String EXTRA_DEVICE_ADDRESS = "device_address";   //Mac 地址
    //定义广播接收者,用于处理扫描蓝牙设备后的结果
    private final BroadcastReceiver mReceiver = new BroadcastReceiver() {
        @Override
        public void onReceive(Context context, Intent intent) {
```

```java
            String action = intent.getAction();
            if (BluetoothDevice.ACTION_FOUND.equals(action)) {
                BluetoothDevice device = intent.getParcelableExtra(BluetoothDevice. EXTRA_
                                                                                    DEVICE);
                if (device.getBondState() != BluetoothDevice.BOND_BONDED) {
                    mNewDevicesArrayAdapter.add(device.getName() + "\n" + device.
                                                                            getAddress());
                }
            } else if (BluetoothAdapter.ACTION_DISCOVERY_FINISHED.equals(action)) {
                if (mNewDevicesArrayAdapter.getCount() == 0) {
                    String noDevices = getResources().getText(R.string.none_found).
                                                                            toString();
                    mNewDevicesArrayAdapter.add(noDevices);
                }
            }
        }
    };
    @Override
    protected void onCreate(Bundle savedInstanceState) {
        super.onCreate(savedInstanceState);
        setContentView(R.layout.device_list);
        //在被调用活动里，设置返回结果码
        setResult(Activity.RESULT_CANCELED);
        init();    //活动界面
    }
    private void init() {
        Button scanButton = findViewById(R.id.button_scan);
        scanButton.setOnClickListener(new OnClickListener() {
            public void onClick(View v) {
                Toast.makeText(DeviceList.this, R.string.scanning, Toast.LENGTH_ LONG).
                                                                            show();
                doDiscovery();    //搜索蓝牙设备
            }
        });
        mPairedDevicesArrayAdapter = new ArrayAdapter<String>(this, R.layout.device_ name);
        mNewDevicesArrayAdapter = new ArrayAdapter<String>(this, R.layout.device_ name);
        //已配对蓝牙设备列表
        ListView pairedListView = findViewById(R.id.paired_devices);
        pairedListView.setAdapter(mPairedDevicesArrayAdapter);
        pairedListView.setOnItemClickListener(mPaireDeviceClickListener);
        //未配对蓝牙设备列表
        ListView newDevicesListView = findViewById(R.id.new_devices);
        newDevicesListView.setAdapter(mNewDevicesArrayAdapter);
        newDevicesListView.setOnItemClickListener(mNewDeviceClickListener);
        //动态注册广播接收者
        IntentFilter filter = new IntentFilter(BluetoothDevice.ACTION_FOUND);
```

```java
            registerReceiver(mReceiver, filter);
            filter = new IntentFilter(BluetoothAdapter.ACTION_DISCOVERY_FINISHED);
            registerReceiver(mReceiver, filter);
            mBtAdapter = BluetoothAdapter.getDefaultAdapter();
            Set<BluetoothDevice> pairedDevices = mBtAdapter.getBondedDevices();
            if (pairedDevices.size() > 0) {
                findViewById(R.id.title_paired_devices).setVisibility(View.VISIBLE);
                for (BluetoothDevice device : pairedDevices) {
                    mPairedDevicesArrayAdapter.add(device.getName() + "\n" + device.
                                                                        getAddress());
                }
            } else {
                String noDevices = getResources().getText(R.string.none_paired).toString();
                mPairedDevicesArrayAdapter.add(noDevices);
            }
    }
    @Override
    protected void onDestroy() {
        super.onDestroy();
        if (mBtAdapter != null) {
            mBtAdapter.cancelDiscovery();
        }
        this.unregisterReceiver(mReceiver);
    }
    private void doDiscovery() {
        findViewById(R.id.title_new_devices).setVisibility(View.VISIBLE);
        if (mBtAdapter.isDiscovering()) {
            mBtAdapter.cancelDiscovery();
        }
        mBtAdapter.startDiscovery();    //开始搜索蓝牙设备并产生广播
         //startDiscovery 是一个异步方法
         //找到一个设备时就发送一个 BluetoothDevice.ACTION_FOUND 的广播
    }
    private OnItemClickListener mPaireDeviceClickListener = new OnItemClickListener() {
        public void onItemClick(AdapterView<?> av, View v, int arg2, long arg3) {
                mBtAdapter.cancelDiscovery();
                String info = ((TextView) v).getText().toString();
                String address = info.substring(info.length() - 17);
                Intent intent = new Intent();
                intent.putExtra(EXTRA_DEVICE_ADDRESS, address);    //Mac 地址
                setResult(Activity.RESULT_OK, intent);
                finish();
        }
    };
    private OnItemClickListener mNewDeviceClickListener = new OnItemClickListener() {
        public void onItemClick(AdapterView<?> av, View v, int arg2, long arg3) {
```

```java
                mBtAdapter.cancelDiscovery();
                Toast.makeText(DeviceList.this, "请在蓝牙设置界面手动连接设备",
                                    Toast.LENGTH_SHORT).show();
                Intent intent = new Intent(Settings.ACTION_BLUETOOTH_SETTINGS);
                startActivityForResult(intent,1);
            }
        };
        //回调方法：进入蓝牙配对设置界面返回后执行
        @Override
        protected void onActivityResult(int requestCode, int resultCode, Intent data) {
            super.onActivityResult(requestCode, resultCode, data);
            init();    //刷新好友列表
        }
}
```

（7）使用菜单 File→Refactor→Rename，重命名模块的 MainActivity 为 BluetoothChat，它是蓝牙会话的主 Activity 组件程序，其文件代码如下：

```java
/*
    本例演示了使用手机蓝牙实现会话功能
    实验步骤：
    (1) 在两个手机分别安装、运行本应用。如果未打开手机蓝牙，则进入打开蓝牙设置界面
    (2) 在 OptionMenu（选项菜单）里，先选择"我的好友"项，即可直接连接，再进行会话
    (3) 新建好友前，需要先使用选项菜单"设置在线"项，即要求对方可见（能被扫描到）。
        然后，单击"搜索好友"按钮
*/
public class BluetoothChat extends AppCompatActivity{
    public static final int MESSAGE_STATE_CHANGE = 1;
    public static final int MESSAGE_READ = 2;
    public static final int MESSAGE_WRITE = 3;
    public static final int MESSAGE_DEVICE_NAME = 4;
    public static final int MESSAGE_TOAST = 5;
    public static final String DEVICE_NAME = "device_name";
    public static final String TOAST = "toast";
    private static final int REQUEST_CONNECT_DEVICE = 1;   //请求连接设备
    private static final int REQUEST_ENABLE_BT = 2;
    private TextView mTitle;
    private ListView mConversationView;
    private EditText mOutEditText;
    private Button mSendButton;
    private String mConnectedDeviceName = null;
    private ArrayAdapter<String> mConversationArrayAdapter;
    private StringBuffer mOutStringBuffer;
    private BluetoothAdapter mBluetoothAdapter = null;
    private ChatService mChatService = null;
    @Override
    public void onCreate(Bundle savedInstanceState) {
```

```java
        super.onCreate(savedInstanceState);
        setContentView(R.layout.activity_main);
        getSupportActionBar().hide();    //隐藏标题栏
        if (Build.VERSION.SDK_INT >= Build.VERSION_CODES.M) {
            if (ContextCompat.checkSelfPermission(this,
                            Manifest.permission.ACCESS_COARSE_LOCATION) !=
                                    PackageManager.PERMISSION_GRANTED) {
                ActivityCompat.requestPermissions(this,new String[]{
                            Manifest.permission.ACCESS_COARSE_LOCATION}, 1);
            }
        }
        Toolbar toolbar = findViewById(R.id.toolbar);
        //创建选项菜单
        toolbar.inflateMenu(R.menu.option_menu);
        //选项菜单监听
        toolbar.setOnMenuItemClickListener(new MyMenuItemClickListener());
        mTitle = findViewById(R.id.title_left_text);
        mTitle.setText(R.string.app_name);
        mTitle = findViewById(R.id.title_right_text);
        // 得到本地蓝牙适配器
        mBluetoothAdapter = BluetoothAdapter.getDefaultAdapter();
        if (mBluetoothAdapter == null) {
            Toast.makeText(this, "蓝牙不可用", Toast.LENGTH_LONG).show();
            finish();
            return;
        }
        if (!mBluetoothAdapter.isEnabled()) { //若当前设备蓝牙功能未开启
            Intent enableIntent = new Intent(BluetoothAdapter.ACTION_REQUEST_ENABLE);
            startActivityForResult(enableIntent, REQUEST_ENABLE_BT); //
        } else {
            if (mChatService == null) {
                setupChat();    //创建会话
            }
        }
    }
    @Override
    public void onRequestPermissionsResult(int requestCode, @NonNull String[] permissions,
                                            @NonNull int[] grantResults) {
        super.onRequestPermissionsResult(requestCode, permissions, grantResults);
        if(grantResults.length>0){
            if(grantResults[0]!=PackageManager.PERMISSION_GRANTED){
                Toast.makeText(this, "未授权,蓝牙搜索功能将不可用! ",
                                            Toast.LENGTH_SHORT).show();
            }
        }
    }
```

```java
@Override
public synchronized void onResume() {    //synchronized：同步方法实现排队调用
    super.onResume();
    if (mChatService != null) {
        if (mChatService.getState() == ChatService.STATE_NONE) {
            mChatService.start();
        }
    }
}
private void setupChat() {
    mConversationArrayAdapter = new ArrayAdapter<String>(this, R.layout.message);
    mConversationView = findViewById(R.id.in);
    mConversationView.setAdapter(mConversationArrayAdapter);
    mOutEditText = findViewById(R.id.edit_text_out);
    mOutEditText.setOnEditorActionListener(mWriteListener);
    mSendButton = findViewById(R.id.button_send);
    mSendButton.setOnClickListener(new OnClickListener() {
        public void onClick(View v) {
            TextView view = findViewById(R.id.edit_text_out);
            String message = view.getText().toString();
            sendMessage(message);
        }
    });
    //创建服务对象
    mChatService = new ChatService(this, mHandler);
    mOutStringBuffer = new StringBuffer("");
}
@Override
public void onDestroy() {
    super.onDestroy();
    if (mChatService != null)
        mChatService.stop();
}
private void ensureDiscoverable() { //修改本机蓝牙设备的可见性
    //打开手机蓝牙后，能被其他蓝牙设备扫描到的时间不是永久的
    if (mBluetoothAdapter.getScanMode() != BluetoothAdapter.
                            SCAN_MODE_CONNECTABLE_DISCOVERABLE) {
        Intent discoverableIntent = new Intent(BluetoothAdapter.
                            ACTION_REQUEST_DISCOVERABLE);
        //设置在 300s 内可见（能被扫描）

discoverableIntent.putExtra(BluetoothAdapter.EXTRA_DISCOVERABLE_DURATION, 300);
        startActivity(discoverableIntent);
        Toast.makeText(this, "已经设置本机蓝牙设备的可见性，对方可搜索了。",
                            Toast.LENGTH_SHORT).show();
    }
```

```java
    }
    private void sendMessage(String message) {
        if (mChatService.getState() != ChatService.STATE_CONNECTED) {
            Toast.makeText(this, R.string.not_connected, Toast.LENGTH_SHORT).show();
            return;
        }
        if (message.length() > 0) {
            byte[] send = message.getBytes();
            mChatService.write(send);
            mOutStringBuffer.setLength(0);
            mOutEditText.setText(mOutStringBuffer);
        }
    }
    private TextView.OnEditorActionListener mWriteListener = new TextView.OnEditorActionListener() {
        @Override
        public boolean onEditorAction(TextView view, int actionId, KeyEvent event) {
            if (actionId == EditorInfo.IME_NULL && event.getAction() == KeyEvent.ACTION_UP) {
                //软键盘里的回车键也能发送消息
                String message = view.getText().toString();
                sendMessage(message);
            }
            return true;
        }
    };
    //使用 Handler 对象在 UI 主线程与子线程之间传递消息
    private final Handler mHandler = new Handler() {      //消息处理
        @Override
        public void handleMessage(Message msg) {
            switch (msg.what) {
            case MESSAGE_STATE_CHANGE:
                switch (msg.arg1) {
                case ChatService.STATE_CONNECTED:
                    mTitle.setText(R.string.title_connected_to);
                    mTitle.append(mConnectedDeviceName);
                    mConversationArrayAdapter.clear();
                    break;
                case ChatService.STATE_CONNECTING:
                    mTitle.setText(R.string.title_connecting);
                    break;
                case ChatService.STATE_LISTEN:
                case ChatService.STATE_NONE:
                    mTitle.setText(R.string.title_not_connected);
                    break;
                }
```

```java
                break;
            case MESSAGE_WRITE:
                byte[] writeBuf = (byte[]) msg.obj;
                String writeMessage = new String(writeBuf);
                mConversationArrayAdapter.add("我:   " + writeMessage);
                break;
            case MESSAGE_READ:
                byte[] readBuf = (byte[]) msg.obj;
                String readMessage = new String(readBuf, 0, msg.arg1);
                mConversationArrayAdapter.add(mConnectedDeviceName + ":  "
                        + readMessage);
                break;
            case MESSAGE_DEVICE_NAME:
                mConnectedDeviceName = msg.getData().getString(DEVICE_NAME);
                Toast.makeText(getApplicationContext(),"链接到 " +
                        mConnectedDeviceName, Toast.LENGTH_SHORT).show();
                break;
            case MESSAGE_TOAST:
                Toast.makeText(getApplicationContext(),
                        msg.getData().getString(TOAST), Toast.LENGTH_SHORT).show();
                break;
            }
        }
    };
    //返回进入好友列表操作后的回调方法
    public void onActivityResult(int requestCode, int resultCode, Intent data) {
        switch (requestCode) {
        case REQUEST_CONNECT_DEVICE:
            if (resultCode == Activity.RESULT_OK) {
                String address = data.getExtras().getString(DeviceList.EXTRA_DEVICE_
                                                                        ADDRESS);
                BluetoothDevice device = mBluetoothAdapter.getRemoteDevice(address);
                mChatService.connect(device);
            }else if(resultCode==Activity.RESULT_CANCELED){
                Toast.makeText(this, "未选择任何好友!", Toast.LENGTH_SHORT). show();
            }
            break;
        case REQUEST_ENABLE_BT:
            if (resultCode == Activity.RESULT_OK) {
                setupChat();
            } else {
                Toast.makeText(this, R.string.bt_not_enabled_leaving, Toast.LENGTH_
                                                                    SHORT).show();
                finish();
            }
        }
```

第9章 Android近距离通信技术

```
         }
    //内部类，选项菜单的单击事件处理
    private class MyMenuItemClickListener implements Toolbar.OnMenuItemClickListener {
        @Override
        public boolean onMenuItemClick(MenuItem item) {
            switch (item.getItemId()) {
                case R.id.scan:
                    //启动DeviceList这个Activity
                    Intent serverIntent = new Intent(BluetoothChat.this, DeviceList. class);
                    startActivityForResult(serverIntent, REQUEST_CONNECT_DEVICE);
                    return true;
                case R.id.discoverable:
                    ensureDiscoverable();
                    return true;
                case R.id.back:
                    finish();
                    System.exit(0);
                    return true;
            }
            return false;
        }
    }
}
```

9.3 近场通信 NFC

9.3.1 NFC 简介

NFC（Near Field Communication）是一种近距离、非接触式识别的无线通信技术，由飞利浦和索尼等公司共同开发，通常有效通信距离在 4cm 以内，通信速率为 106～848 Kb/s，工作频率为 13.56 MHz。

通过 NFC 技术，可以实现 Android 设备与 NFC Tag（Target）或其他 Android 设备之间小批量数据的传输。NFC 可以在移动设备、消费类电子产品、PC 和智能控件工具间进行近距离无线通信。

注意：NFC 从 RFID（Radio Frequency Identification Device，无线射频识别装置）发展而来，与 RFID 没有太大区别，都是基于地理位置相近的两个物体之间的信号传输。

NFC 通信总是由一个发起者（Initiator）和一个被动式接受者（Passive Target）组成。通常发起者主动发送电磁场可以为被动式接受者提供电源。因此，被动式接受者（Tag）可以有非常简单的形式，如标签、芯片卡和纽扣等。

具有 NFC 功能的手机与 Tag 如图 9.3.1 所示。

注意：具有 NFC 功能的手机，如小米手机，除了能读/写 Tag 标签外，还可以作为公交卡使用，这说明它本身含有 Tag。

图 9.3.1　具有 NFC 功能的手机与 Tag

NFC 芯片作为组成 RFID 模块的一部分装在手机上，手机背后有一块 NFC 的区域，如图 9.3.2 所示。

图 9.3.2　手机背后的 NFC 区域

在 Android NFC 应用中，Android 手机通常是通信中的发起者和 NFC 的读写器。Android 手机也可以模拟 NFC 通信的接收者，从而实现点到点的通信（P2P）。

当 Tag 接近手机 NFC 区域时，会产生提示音，同时以对话框形式呈现手机所有的 NFC 应用程序，等待用户选择，如图 9.3.3 所示。

图 9.3.3　Tag 靠近手机 NFC 区域时产生的对话框

和其他无线通信方式（如 Bluetooth）相比，NFC 支持的通信带宽和距离要小得多，但是标签成本低，也不需要搜寻设备、配对等，通信在靠近的瞬间即可完成。

9.3.2 Android 对 NFC 的支持

Android 对 NFC 的支持主要在 android.nfc 和 android.nfc.tech 两个包中，如图 9.3.4 所示。

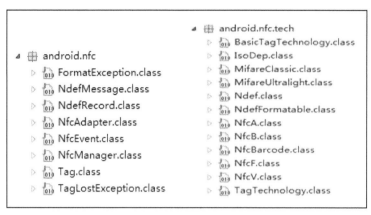

图 9.3.4 Android 提供的 NFC 软件包

在手机的 NFC 应用中，需要在清单文件中添加 NFC 权限及设备特征，其代码如下：

```
<uses-permission android:name="android.permission.NFC" />
<!--从 Google 应用商店下载安装 NFC 应用时,没有 NFC 功能的手机将不被安装-->
<uses-feature
    android:name="android.hardware.nfc"
    android:required="true" />
```

NFCManager 类可以用来管理 Android 设备中指出的所有 NFCAdapter，但由于大部分 Android 设备只支持一个 NFCAdapter，可使用该类提供的 getDefaultAdapter() 来获取系统支持的 NFCAdapter。

注意：NFCAdapter 类也提供了静态方法 getDefaultAdapter()来获取 NFCAdapter 对象。

当 Android 设备检测到一个 Tag 时，会创建一个 Tag 对象，将其放在 Intent 对象，然后发送到相应的 Activity。

NFCAdapter 类可以用来定义一个 Intent，在系统检测到 NFC Tag 时通知事先定义的 Activity，以实现对 Tag 设备的读/写操作。

类 NdefMessage 和 NdefRecord 是 NFC forum 定义的数据格式。

在包 android.nfc.tech 中，定义了可以对 Tag 设备进行读/写操作的类。

9.3.3 NFC 应用实例：读/写 Tag 标签

下面介绍读/写 Tag 标签的程序设计。

【例 9.3.1】 使用手机 NFC 功能读/写 Tag 标签。

模块运行时，默认执行读 Tag 功能。单击"写入标签"按钮并输入要写入的文本后，再次单击"写入标签"按钮时的运行效果，如图 9.3.5 所示。

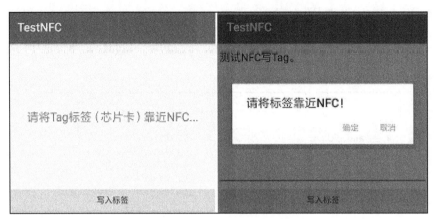

图 9.3.5　NFC 读/写模块的运行界面

主要设计步骤如下：

（1）新建名为 exaple9_2 的模块。在模块清单文件里，添加注册本应用所需要的相关权限。

（2）编写 Activity 组件 ReadTag，它没有包含危险权限，调用 Activity 组件 WriteTag，其文件代码如下：

```
/*
    本活动用于读出 Tag 标签里的内容：非接触式，NFC 线圈靠近芯片卡的中央
    如果读取其他已经加密的卡，程序会闪退
*/
public class ReadTag extends AppCompatActivity {
    private NfcAdapter nfcAdapter;
    private TextView resultText;
    private PendingIntent pendingIntent;
    private IntentFilter[] mFilters;
    private String[][] mTechLists;
    private Button mJumpTagBtn;
    private boolean isFirst = true;
    @Override
    protected void onCreate(Bundle savedInstanceState) {
        super.onCreate(savedInstanceState);
        // 获取 NFC 适配器，判断设备是否支持 NFC 功能
        nfcAdapter = NfcAdapter.getDefaultAdapter(this);
        if (nfcAdapter == null) {
            //提示文本在资源文件 res/values/strings.xml 里，R.string.no_nfc 是文本资源的 Id 值
            Toast.makeText(this, getResources().getString(R.string.no_nfc),
                                                Toast.LENGTH_SHORT).show();
```

```java
        finish();
        return;
    } else if (!nfcAdapter.isEnabled()) {
        //NFC 未打开时
        Toast.makeText(this, getResources().getString(R.string.open_nfc),
                                        Toast.LENGTH_SHORT).show();
        finish();
        return;
    }
    setRequestedOrientation(ActivityInfo.SCREEN_ORIENTATION_PORTRAIT); //设定竖屏
    setContentView(R.layout.read_tag); //进入 ReadTag 活动，在清单文件里配置作为主活动
    resultText = findViewById(R.id.resultText); // 显示结果 Text
    mJumpTagBtn = findViewById(R.id.jump); // 写入标签按钮
    mJumpTagBtn.setOnClickListener(new OnClickListener() {
        @Override
        public void onClick(View v) {
            switch (v.getId()) {
                case R.id.jump:
                    Intent intent = new Intent(ReadTag.this, WriteTag.class);
                    startActivity(intent);   //调用 WiteTag 这个 Activity
                default:
                    break;
            }
        }
    });
    /*
        创建一个 PendingIntent 对象，在扫描到 NFC 标签时，用它来封装 NFC 标签的详
                                                            细信息。
        Intent.FLAG_ACTIVITY_SINGLE_TOP：如果当前栈顶的 activity 就是要启动的
                                                                    activity,
        则不会再启动一个新的 activity
    */
    pendingIntent = PendingIntent.getActivity(this, 0, new Intent(this,
                    getClass()).addFlags(Intent.FLAG_ACTIVITY_SINGLE_TOP), 0);
    //使用 IntentFilter 过滤想要的 action
    IntentFilter ndef = new IntentFilter(NfcAdapter.ACTION_TECH_DISCOVERED);
    ndef.addCategory("*/*");
    mFilters = new IntentFilter[] { ndef };     //过滤器
    mTechLists = new String[][] {
                //如果 android 设备支持 MIFARE，提供对 MIFARE Classic 目标的属性和 I/O
                                                                        操作
                new String[] { MifareClassic.class.getName() },
                new String[] { NfcA.class.getName() } };    // 允许扫描的标签类型
}
//重写 Activity 组件的 onNewIntent()方法（不是生命周期方法）来获取扫描到的 NFC 标签数据
```

```java
@Override
protected void onNewIntent(Intent intent) {    //配合延期意图使用
    super.onNewIntent(intent);
    /*当系统检测到 Tag 标签中含有 NDEF 格式的数据时，且系统中有 Activity 声明可以接
      受包含 NDEF 数据的 Intent 的时候，系统会优先发出这个 action 的 intent*/
    if (NfcAdapter.ACTION_TECH_DISCOVERED.equals(intent.getAction())) {
        String result = processIntent(intent);
        resultText.setText(result);
    }
}
@Override
protected void onResume() {    //Activity 组件的生命周期方法，配合延期意图使用
    super.onResume();
    nfcAdapter.enableForegroundDispatch(this, pendingIntent, mFilters,mTechLists);
    if (isFirst) {
        if (NfcAdapter.ACTION_TECH_DISCOVERED.equals(getIntent().getAction())) {
            String result = processIntent(getIntent());
            resultText.setText(result);
        }
        isFirst = false;
    }
}
//获取标签中的内容
private String processIntent(Intent intent) {
    String resultStr=getResources().getString(R.string.message);
    try {
        Parcelable[] rawmsgs = intent.getParcelableArrayExtra(
                              NfcAdapter.EXTRA_NDEF_MESSAGES);
        NdefMessage msg = (NdefMessage) rawmsgs[0];
        NdefRecord[] records = msg.getRecords();
        resultStr = new String(records[0].getPayload());
    }catch (Exception e){
        Toast.makeText(this, "无法识别的卡！", Toast.LENGTH_SHORT).show();
    }
    return resultStr;
}
}
```

（3）写 Tag 标签 Activity 组件 WriteTag 的文件代码如下：

```java
//本 Activity 组件 WriteTag 供本 Activity 组件 ReadTag 调用
public class WriteTag extends AppCompatActivity {
    private IntentFilter[] mWriteTagFilters;
    private NfcAdapter nfcAdapter;
    PendingIntent pendingIntent;
    String[][] mTechLists;
    Button writeBtn;
```

```java
boolean isWrite = false;
EditText mContentEditText;
@Override
protected void onCreate(Bundle savedInstanceState) {
    super.onCreate(savedInstanceState);
    setRequestedOrientation(ActivityInfo.SCREEN_ORIENTATION_PORTRAIT); //设定竖屏
    setContentView(R.layout.write_tag);
    writeBtn = findViewById(R.id.writeBtn);
    writeBtn.setOnClickListener(new OnClickListener() {
        @Override
        public void onClick(View v) {
            isWrite = true;
            AlertDialog.Builder builder = new AlertDialog.Builder(WriteTag.this).
                                                setTitle("请将标签靠近！");
            builder.setNegativeButton("确定", new DialogInterface.OnClickListener() {
                @Override
                public void onClick(DialogInterface dialog, int which) {
                    dialog.dismiss();
                    mContentEditText.setText("");
                    isWrite = false;
                    WriteTag.this.finish();
                }
            });
            builder.setPositiveButton("取消", new DialogInterface.OnClickListener() {
                @Override
                public void onClick(DialogInterface dialog, int which) {
                    dialog.dismiss();
                    isWrite = false;
                }
            });
            builder.create();
            builder.show();
        }
    });
    mContentEditText = findViewById(R.id.content_edit);
    nfcAdapter = NfcAdapter.getDefaultAdapter(this);
    if (nfcAdapter == null) {
        Toast.makeText(this, getResources().getString(R.string.no_nfc),
                                    Toast.LENGTH_SHORT).show();
        finish();
        return;
    } else if (!nfcAdapter.isEnabled()) {
        Toast.makeText(this, getResources().getString(R.string.open_nfc),
                    Toast.LENGTH_SHORT).show();
        finish();
        return;
```

```java
            }
            pendingIntent = PendingIntent.getActivity(this, 0, new Intent(this,
                    getClass()).addFlags(Intent.FLAG_ACTIVITY_SINGLE_TOP), 0);
            IntentFilter writeFilter = new IntentFilter(NfcAdapter.ACTION_TECH_DISCOVERED);
            mWriteTagFilters = new IntentFilter[] { writeFilter };
            mTechLists = new String[][] {
                    new String[] { MifareClassic.class.getName() },
                    new String[] { NfcA.class.getName() } };// 允许扫描的标签类型
    }
    //重写 Activity 组件的 onNewIntent()方法（不是生命周期方法）来获取扫描到的 NFC 标签
                                                                            数据
    @Override
    protected void onNewIntent(Intent intent) {    //配合延期意图使用
        super.onNewIntent(intent);
        if (isWrite == true && NfcAdapter.ACTION_TECH_DISCOVERED.equals(intent.
                                                                    getAction())) {
            Tag tag = intent.getParcelableExtra(NfcAdapter.EXTRA_TAG);
            NdefMessage ndefMessage = getNoteAsNdef();
            if (ndefMessage != null) {
                writeTag(getNoteAsNdef(), tag);
            } else {
                showToast("请输入您要写入标签的内容");
            }
        }
    }
    @Override
    protected void onResume() {    //Activity 组件的生命周期方法，配合延期意图使用
        super.onResume();
        nfcAdapter.enableForegroundDispatch(this, pendingIntent,
                mWriteTagFilters, mTechLists);
    }
    // 根据文本生成一个 NdefRecord
    private NdefMessage getNoteAsNdef() {
        String text = mContentEditText.getText().toString();
        if (text.equals("")) {
            return null;
        } else {
            byte[] textBytes = text.getBytes();
            NdefRecord textRecord = new NdefRecord(NdefRecord.TNF_MIME_MEDIA,
                    "image/jpeg".getBytes(), new byte[] {}, textBytes);
            return new NdefMessage(new NdefRecord[] { textRecord });
        }
    }
    // 写入 Tag 标签
    boolean writeTag(NdefMessage message, Tag tag) {
        int size = message.toByteArray().length;
```

```java
            try {
                Ndef ndef = Ndef.get(tag);
                if (ndef != null) {
                    ndef.connect();
                    if (!ndef.isWritable()) {
                        showToast("Tag 不允许写入");
                        return false;
                    }
                    if (ndef.getMaxSize() < size) {
                        showToast("文件大小超出容量");
                        return false;
                    }
                    ndef.writeNdefMessage(message);
                    showToast("写入数据成功.");
                    return true;
                } else {
                    NdefFormatable format = NdefFormatable.get(tag);
                    if (format != null) {
                        try {
                            format.connect();
                            format.format(message);
                            showToast("格式化并且写入 message");
                            return true;
                        } catch (IOException e) {
                            showToast("无法识别的卡！");
                            return false;
                        }
                    } else {
                        showToast("Tag 不支持 NDEF");
                        return false;
                    }
                }
            } catch (Exception e) {
                showToast("写入数据失败");
            }
            return false;
        }
        private void showToast(String text) {
            Toast.makeText(this, text, Toast.LENGTH_SHORT).show();
        }
    }
```

习 题 9

一、判断题

1．没有插入 SIM 卡的手机，不能通过 WiFi 上网。
2．蓝牙通信使用 HTTP 协议。
3．两台蓝牙手机进行文件传输前，必须经过配对。
4．蓝牙设备配对时，会产生广播、动态注册和使用广播接收者程序。
5．文件的蓝牙传输，只能在两个手机之间进行。
6．蓝牙聊天程序设计必须使用多线程。
7．如果手机配置中有 NFC，则在"设置"程序中可以找到。

二、选择题

1．下列通信中，对距离没有限制的是____。
　　A．WiFi　　　　B．NFC　　　　C．GPRS　　　　D．Bluetooth
2．获取手机 WiFi 连接无线路由器的物理地址，应使用 WifiInfo 类的____方法。
　　A．getMacAddress()　　　　B．getSSID()
　　C．getBSSID()　　　　　　　D．getIpAddress()
3．获取已经配对的蓝牙设备列表，需要使用____类提供的方法 getBondedDevice()。
　　A．Bluetooth　　　　　　　　B．BluetoothManager
　　C．BluetoothDevice　　　　　D．BluetoothAdapter
4．下列选项中，涉及 Internet 网络的是____。
　　A．WiFi　　　　B．GPS　　　　C．Bluetooth　　　D．NFC
5．下列不属于 android.bluetooth 包的选项是____。
　　A．BluetoothAdapter　　　　　B．BluetoothSocket
　　C．BluetoothServerSocket　　　D．UUID

三、填空题

1．WiFi 状态共有____种。
2．为了获取扫描到的所有 WiFi 信息，在使用 WifiManager 的 getScanResults()方法前，应要使用 WifiManager 的____方法。
3．获取手机当前连接 WiFi 信源的相关信息，需要使用 WifiManager 的____方法。
4．为了得到本地蓝牙适配器，需要调用 BluetoothAdapter 类的静态方法____。
5．创建蓝牙 RFID 连接方法 createRfcommSocketToServiceRecord()的参数类型是____。

第9章 Android 近距离通信技术

实 验 9

一、实验目的

(1) 掌握手机 WiFi 的使用及 WiFi 程序设计。
(2) 掌握手机的蓝牙功能及其程序设计。
(3) 掌握 Android 手机 NFC 功能及其程序设计。

二、实验内容及步骤

【预备】在 Android Studio 中，新建名为 Example9 的项目后，访问本课程配套的网站 http://www.wustwzx.com/as/sy/sy09.html，复制相关代码，完成如下几个模块的设计。

1. 掌握手机 WiFi 的使用和 WiFi 功能的程序实现

(1) 运行手机设置程序，熟悉 WiFi 的打开、关闭、扫描、选择等操作。

(2) 打开手机的移动数据连接，运行手机自带的"网络分享和便携式热点"程序，选择 WiFi 连接方式，共享手机的移动数据连接。

(3) 在项目 Example 9 里，新建名为 example9_1 的模块，在其清单文件里，注册 CHANGE_WIFI_STATE 和 ACCESS_WIFI_STATE 两个普通权限，注册 1 个 WiFi 扫描所需的危险权限 ACCESS_COARSE_LOCATION。

(4) 在布局文件里，分别添加 4 个用于 WiFi 操作的 Button 按钮，1 个用于产生卷动效果的 ScrollView 控件并内嵌入 1 个 TextView 控件。

(5) 在 MainActivity 的 onCreate()方法里，动态申请危险权限，并重写请求权限的回调方法 onRequestPermissionsResult()，分别实现 4 个按钮功能的实现代码，将 WiFi 扫描的结果显示在 TextView 控件。

(6) 部署本模块，做运行测试。

2. 掌握手机蓝牙功能的使用和 Android 蓝牙聊天程序设计

(1) 运行手机设置程序，熟悉蓝牙的打开、关闭、扫描和配对等操作。

(2) 在已经配对成功的两个蓝牙设备之间实现文件的蓝牙传输。

(3) 在项目 Example 9 里，新建名为 example9_2 的模块，在其清单文件里，注册 BLUETOOTH_ ADMIN 和 BLUETOOTH 两个普通权限,注册 1 个蓝牙设备搜索所需的危险权限 ACCESS_COARSE_LOCATION。

(4) 在 res/menu 文件夹里，创建供主 Activity 组件使用的菜单文件。

(5) 编写用于蓝牙会话的服务组件 ChatService，创建蓝牙连接请求、连接管理和通信服务等线程类作为内部类。

(6) 新建 Activity 组件 DeviceList，搜索可配对的好友进行配对，显示已配对的好友列表，选取与之会话的蓝牙设备。

（7）使用菜单 File→Refactor→Rename，重命名模块的 MainActivity 为 BluetoothChat，它是蓝牙会话的主 Activity 程序。新建好友前，需要先使用选项菜单"设置在线"项，以供蓝牙搜索。选择好友，创建服务对象，调用服务，实现蓝牙会话功能。

（8）分别部署本模块到两部手机上运行，对其中一部手机使用菜单键选择与之会话的另一部手机，建立好连接后做会话测试。

3．掌握 Android 手机的 NFC 功能及程序设计

（1）运行手机设置程序，熟悉 NFC 的打开、关闭及测试等操作。

（2）在项目 Example 9 里，新建名为 example9_3 的模块，在其清单文件里，注册使用 NFC 的普通权限。

（3）使用菜单 File→Refactor→Rename，重命名模块的 MainActivity 为 ReadTag，在其 onCreate()方法里，检测设备是否具有 NFC 功能、创建 NFCAdapter 对象。

（4）使用 PendingIntent 和 IntentFilter，配合重写 Activity 组件方法 onNewIntent()及生命周期方法 onResume()获取 Tag 标签数据，并刷新 TextView 控件。

（5）在模块中新建 Activity 组件 WriteTag，使用 PendingIntent 和 IntentFilter，配合重写 Activity 组件方法 onNewIntent()及生命周期方法 onResume()，将 EditText 控件里的数据写入 Tag 标签。

（6）部署本模块，做运行测试。

三、实验小结及思考

(由学生填写，重点写上机中遇到的问题。)

第 10 章　位置服务与地图应用开发

位置服务（Location Based Services，LBS）又称定位服务，是指通过 GPS 卫星或者网络（WiFi 或 GPRS）获取各种终端的地理坐标（经度和纬度），在电子地图平台的支撑下，为用户提供基于位置导航、查询的一种信息服务。位置服务是 Android 设备的一个重要功能，本章学习要点如下：

- Android 提供的 GPS 定位系统服务及相关类；
- Android 提供的 WiFi 定位系统服务及相关类；
- 百度提供定位服务的相关类及其方法；
- 百度地图应用。

10.1　位置服务概述

10.1.1　基于位置的服务 LBS

位置服务是移动设备的一个重要功能，Android 提供了基于位置的服务 LBS。在手机设置程序里，打开定位服务时，有如下定位模式可供选择。

- 高精度定位模式：这种定位模式会同时使用网络定位和 GPS 定位，优先返回最高精度的定位结果；
- 低功耗定位模式：这种定位模式不会使用 GPS 进行定位，只会使用网络定位（WiFi 定位和 GPRS 基站定位）；
- 仅用设备定位模式：这种定位模式不需要连接网络，只使用 GPS 进行定位。它不支持室内环境的定位。

1．GPS 定位

GPS 是全球定位系统（Global Positioning System）的英文缩写，是 20 世纪 70 年代由美国陆海空三军联合研制的新一代空间卫星导航系统。基于 GPS 的定位方式是利用手机上的 GPS 定位模块（芯片）与空中的 GPS 卫星之间的通信实现的。由于卫星的位置精确，在 GPS 观测中，可以得到卫星到接收机之间的距离，然后利用三维坐标中的距离公式和 3 颗卫星就可以组成 3 个方程式，解出观测点的位置坐标（x, y, z）。考虑到卫星时钟与接收机时钟之间的误差，因此，至少需要 4 颗卫星信号才能进行 GPS 定位。

注意：

（1）手机 GPS 信号在户外空旷地方一般正常；而在室内或在地下通道内，手机可能没有 GPS 信号。

（2）开启 Android 设备到 GPS 信号需要一定的时间，室内靠近窗户的地方大约需要 2 分钟，而室外很快。

（3）相对于下面介绍的两种网络定位方式（WiFi 定位和基站定位），GPS 定位具有较高的精度。

2．WiFi 定位

WiFi 是 Wireless Fidelity 的英文缩写，表示无线相容认证。

WiFi 能够对用户进行定位，因为在 Android、iOS 和 Windows Phone 这些手机操作系统中内置了位置服务。由于每一个 WiFi 热点都有一个独一无二的 MAC 地址（手机 WiFi 网卡的 MAC 地址），智能手机开启 WiFi 后就会自动扫描附近热点并上传其位置信息，这样就建立了一个庞大的热点位置数据库，这个数据库是对用户进行定位的关键。

注意：

（1）WiFi 定位和 GPRS 定位，统称为网络定位。

（2）网络定位响应速度快、费电少，但精度没有 GPS 定位高。

3．GPRS 基站定位

基站定位是通过移动数据中的 GPRS （General Packet Radio Service，通用分组无线服务技术）网络，利用基站对手机的距离测算来确定手机位置的。因此，基站定位也称 GPRS 定位。

GPRS 是一种基于 GSM 系统的无线分组交换技术，提供了端到端的、广域的无线 IP 连接。基站定位服务一般应用于手机用户，手机基站定位服务是通过网络运营商（主要有中国电信、中国移动和中国联通）的移动网络获取移动终端用户的位置信息（经度和纬度坐标），在电子地图平台的支持下，为用户提供相应服务的一种增值业务。

10.1.2 Android API 提供的位置包

手机定位是指能得到手机持有人所在地理位置的经度和纬度。Android 系统提供了用于计算地理数据的相关软件包（类）。Android API 的位置包 android.location，如图 10.1.1 所示。

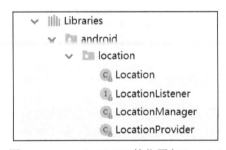

图 10.1.1　Android API 的位置包 location

为了实现定位信息的实时显示，需要使用位置管理器类 LocationManager 的位置请求更新方法 requestLocationUpdates()和位置监听器接口 LocationListener，如图 10.1.2 所示。

242

第 10 章 位置服务与地图应用开发

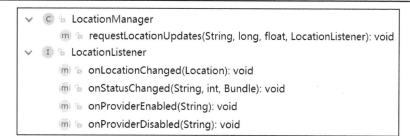

图 10.1.2 位置请求更新方法与位置监听接口

方法 requestLocationUpdates()的 4 个参数含义如下：
- 第一个参数为位置提供者类型；
- 第二个参数设定多久更新一次位置，以 ms 为单位；
- 第三个参数设定位置变化至少多少才更新，以 m 为单位；
- 第四个参数为处理位置变化的监听器对象。

位置监听器接口需要实现的方法如下：

```
@Override
public void onLocationChanged(Location location) {
    // TODO Auto-generated method stub
}
@Override
public void onProviderDisabled(String provider) {
    // TODO Auto-generated method stub
}
@Override
public void onProviderEnabled(String provider) {
    // TODO Auto-generated method stub
}
@Override
public void onStatusChanged(SString provider, int status, Bundle extras) {
    // TODO Auto-generated method stub
}
```

10.1.3 Google Map APIs

Google Map APIs 是指在标准 Android API 的基础上增加了用于开发地图应用的 Map API。Google Map APIs 定义了一系列用于在 Google Map 上显示、控制和层叠信息的功能类。

Google Map APIs 不属于标准 Android SDK 的标准库组件，需要单独下载，可使用 Android SDK Manager 下载，如图 10.1.3 所示。

新建地图应用项目时，Google Map APIs 与标准 Android 项目操作的不同在于编译 API 的选取上。

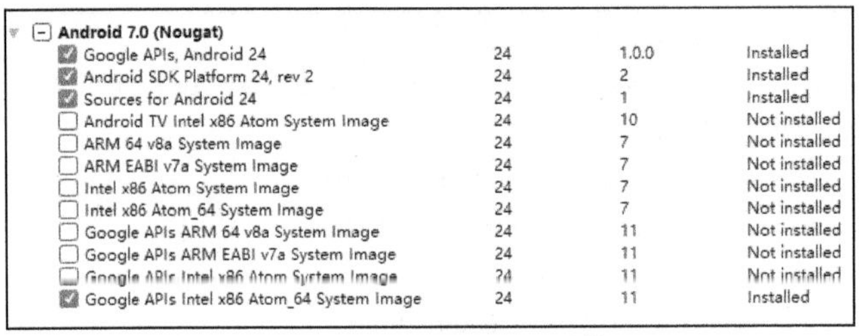

图 10.1.3 使用 Android SDK Manager 下载 Google Map APIs

10.2 Android 定位实现

Android 6.0 及以上版本的定位，首先需要在清单文件里注册定位权限。其次，需要在 Activity 组件的 onCreate()方法里动态申请危险权限 ACCESS_FINE_LOCATION（用于 GPS 定位）或 ACCESS_COARSE_LOCATION（用于网络定位）。由于这两个权限处于同一个权限组，因此，只需在 onCreate()方法里请求其中的一个，其代码框架如下：

```
//如果没有 ACCESS_FINE_LOCATION 权限，动态请求用户允许使用该权限
if (ActivityCompat.checkSelfPermission(this, Manifest.permission. ACCESS_FINE_ LOCATION) != PackageManager.PERMISSION_GRANTED) {
    //请求权限；为了让权限授予时立即生效，一般需要重写权限回调方法 onRequestPermissionsResult()
    //第一个参数为上下文对象，第二个参数为请求的权限，第三个参数为请求码（权限回调方法需要）
    ActivityCompat.requestPermissions(this, new String[]
            {Manifest.permission. ACCESS_FINE_LOCATION }, 1);
} else {
    //调用自定义的定位方法
}
```

在 Activity 组件里，写权限的回调方法，其代码如下：

```
@Override
public void onRequestPermissionsResult(int requestCode, @NonNull String[] permissions,
                                        @NonNull int[] grantResults) {
    super.onRequestPermissionsResult(requestCode, permissions, grantResults);
    switch (requestCode) {
        case 1:
            if (grantResults[0] == PackageManager.PERMISSION_GRANTED){//权限授予时
                //调用自定义的定位方法
            } else {
                Toast.makeText(this, "没有授予定位权限！", Toast.LENGTH_ LONG).
                                                                        show();
                finish();    //退出
            }
    }
}
```

第 10 章 位置服务与地图应用开发

在开始定位前,需要检查手机的定位服务是否打开。打开手机定位服务时,只能从三种定位模式(混合定位、网络定位和仅 GPS 定位)里选择其中一种。因此,检测是否打开定位服务的代码如下:

```
LocationManager lm=getSystemService(Context.LOCATION_SERVICE);
if (lm.isProviderEnabled(LocationManager.GPS_PROVIDER) ||
                          lm.isProviderEnabled(LocationManager.NETWORK_PROVIDER)) {
    //调用自定义的定位方法,此方法在其他回调方法里还会使用
}else{
    //创建进入定位服务设置程序的意图对象
    Intent intent = new Intent(Settings.ACTION_LOCATION_SOURCE_SETTINGS);
    //进入定位服务设置程序,有返回值的调用,第二个参数为请求码
    startActivityForResult(intent, 1);
}
```

用户进入设置程序打开定位服务后,为了在应用程序里立即生效,还必须重写与 startActivityForResult ()方法相对应的回调方法 onActivityResult(),其代码如下:

```
//下面的方法,是重写的 Activity 组件方法,对应于方法 startActivityForResult()
@Override
protected void onActivityResult(int requestCode, int resultCode, @Nullable Intent data) {
    super.onActivityResult(requestCode, resultCode, data);
    if (requestCode == 1) {    //假定进入打开定位服务程序的请求码为 1
        //调用自定义的定位方法;
    }
}
```

10.2.1 GPS 定位实现

开发 Android 6.0 及以上版本的 GPS 定位应用,除了要在清单文件中注册定位权限外,还必须在 Activity 组件里动态申请危险权限 ACCESS_FINE_LOCATION。在清单文件里,注册权限的代码如下:

```
<uses-permission android:name="android.permission.ACCESS_FINE_LOCATION" />
```

编写 Android GPS 实时定位程序,需要使用位置管理器类 LocationManager、位置监听器接口 LocationListener 和位置类 Location,其示例代码如下:

```
//动态申请危险权限 ACCESS_FINE_LOCATION 的代码写在 Activity 组件的 onCreate()方法里略
//申请动态权限的回调方法(略)
//下面的代码,通常封装在 Activity 组件的一个方法里,被有多处调用
//若 GPS 未开启
if(!lm.isProviderEnabled(LocationManager.GPS_PROVIDER))
    //Toast.makeText(currentActivity, "请开启 GPS! ", Toast.LENGTH_SHORT).show();
    //开启设置 GPS 的界面
    Intent intent = new Intent(Settings.ACTION_LOCATION_SOURCE_SETTINGS);
    startActivity(intent);
```

```
        //后面有设置完成后的回调处理方法
}
LocationManager lm=(LocationManager)getSystemService(Context.LOCATION_SERVICE);
if(!lm.isProviderEnabled(Lm.GPS_PROVIDER)){
        Toast.makeText(MainActivity.this,"请开启 GPS 服务",Toast.LENGTH_LONG).show();
        Intent intent = new Intent(Settings.ACTION_LOCATION_SOURCE_SETTINGS);
        startActivityForResult (intent,1);     //第 2 个参数为请求码
}
//Android 6.0 请求位置更新前,需要检查危险权限 ACCESS_FINE_LOCATION
if (ActivityCompat.checkSelfPermission(this, Manifest.permission.ACCESS_FINE_
                                                                LOCATION) !=
                                        PackageManager.PERMISSION_GRANTED) {
        return;
}
lm.requestLocationUpdates(LocationManager.GPS_PROVIDER, 1000, 1, new Location Listener() {
        @Override
        public void onStatusChanged(String arg0, int arg1, Bundle arg2) {
                // TODO Auto-generated method stub
        }
        @Override
        public void onProviderEnabled(String arg0) {
                // TODO Auto-generated method stub
        }
        @Override
        public void onProviderDisabled(String arg0) {
                // TODO Auto-generated method stub
        }
        @Override
        public void onLocationChanged(Location location) {
                // TODO Auto-generated method stub
                String string = "纬度为:" + location.getLatitude() + "\n 经度为:" + location.
                                                                        getLongitude();
                tv.setText(string);    //刷新 UI 控件
        }
});
//下面的方法,是重写的 Activity 组件方法,对应于方法 startActivityForResult()
@Override
protected void onActivityResult(int requestCode, int resultCode, @Nullable Intent data) {
        super.onActivityResult(requestCode, resultCode, data);
        if (requestCode == 1) {   //进入定位设置的请求码在前面设置为 1
                //调用自定义的定位方法;
        }
}
```

注意:在使用方法 requestLocationUpdates()实现位置实时更新之前,必须检查危险权限 ACCESS_FINE_LOCATION,尽管在 onCreate()方法里已经检查和请求了该权限。

10.2.2 网络连接及状态相关类

前面介绍的 GPS 定位方式不需要使用网络连接，但是 WiFi 定位和 GPRS 定位这两种方式需要使用网络连接，因此，这里先介绍与网络连接相关的类。

Android 软件包 android.net 里，提供了与网络连接相关的类，用于网络定位时判定是否有连接、连接状态等，如图 10.2.1 所示。

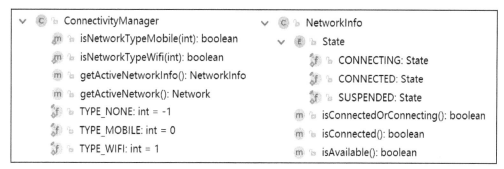

图 10.2.1　Android 网络连接相关类的定义

检查是否有网络连接（GPRS 或 WiFi）的代码如下：

```
ConnectivityManager cm =(ConnectivityManager)getSystemService(Context. CONNECTIVITY_SERVICE);
NetworkInfo networkInfo = cm.getActiveNetworkInfo();
if (networkInfo==null)
    Toast.makeText(this,"没有网络连接",Toast.LENGTH_LONG).show();
else
    Toast.makeText(this,"有网络连接",Toast.LENGTH_LONG).show();
```

检查是否有 GPRS 或 WiFi 连接的代码如下：

```
ConnectivityManager cm =(ConnectivityManager)getSystemService(Context.CONNECTIVITY_SERVICE);
NetworkInfo.State mobileState=cm.getNetworkInfo(ConnectivityManager.TYPE_MOBILE).getState();
if (mobileState==State.CONNECTED||mobileState==State.CONNECTING)
    Toast.makeText(this,"可以使用移动网络",Toast.LENGTH_LONG).show();
else
    Toast.makeText(this,"移动网络未开启!",Toast.LENGTH_LONG).show();
NetworkInfo.State wifiState=cm.getNetworkInfo(ConnectivityManager.TYPE_WIFI).getState();
if (wifiState==State.CONNECTED||wifiState==State.CONNECTING)
    Toast.makeText(this,"可以使用 WiFi 网络!", Toast.LENGTH_LONG) .show();
else
    Toast.makeText(this," 没有 WiFi 网络！ ",Toast.LENGTH_LONG).show();
```

10.2.3　WiFi 或 GPRS 定位实现

使用网络定位，即使用 WiFi 或 GPRS 定位，需要在清单文件里添加如下权限：

```xml
<!--定位权限 ACCESS_COARSE_LOCATION 是危险权限并处于一个权限组,需要动态申请 -->
<!--粗糙定位-->
<uses-permission android:name="android.permission.ACCESS_COARSE_LOCATION"/>
<!—下面的 2 个权限是普通权限,不需要动态申请 -->
<!--网络状态-->
<uses-permission android:name="android.permission.ACCESS_NETWORK_STATE"/>
<!--WiFi 状态-->
<uses-permission android:name="android.permission.ACCESS_WIFI_STATE"/>
```

在 Activity 组件里,网络定位的实现代码如下:

```java
//得到一个位置管理器对象
LocationManager lm=getSystemService(Context.LOCATION_SERVICE);
//得到 WiFi 管理器对象 wfm
WifiManager wfm=getSystemService(Context.WIFI_SERVICE);
//得到位置管理器对象
ConnectivityManager cm = (ConnectivityManager) getSystemService(Context. CONNECTIVITY_SERVICE);
if (cm.getActiveNetworkInfo() != null) { //有网络连接时
    if (wfm.isWifiEnabled()) { //WIFI 有效时,优先使用
        Toast.makeText(getApplicationContext(), "WiFi 方式的网络定位", Toast.LENGTH_LONG).show();
    } else {
        Toast.makeText(getApplicationContext(), "基站方式的网络定位", Toast.LENGTH_LONG).show();
    }
} else { //没有网络连接时
    new AlertDialog.Builder(this)
        .setIcon(R.mipmap.ic_launcher)
        .setTitle("消息框")
        .setMessage("无网络连接!请先设置一种网络连接...")
        .setPositiveButton("确定", new DialogInterface.OnClickListener() {
            @Override
            public void onClick(DialogInterface dialog, int which) {
                //进入网络连接设置
                Intent intent = new Intent(Settings.ACTION_SETTINGS);
                //有返回值的调用,设置网络请求码为 2
                startActivityForResult(intent, 2);
            }
        })
        .show();
}
//请求位置实时更新
if (ActivityCompat.checkSelfPermission(this, Manifest.permission.ACCESS_COARSE_ LOCATION)
                                != PackageManager.PERMISSION_GRANTED) {
    return;
}
```

```
lm.requestLocationUpdates(LocationManager.NETWORK_PROVIDER, 1000, 1,
                                     new LocationListener() {      //实时监听
    @Override
    public void onStatusChanged(String arg0, int arg1, Bundle arg2) {
        // TODO Auto-generated method stub
    }
    @Override
    public void onProviderEnabled(String arg0) {
        // TODO Auto-generated method stub
    }
    @Override
    public void onProviderDisabled(String arg0) {
        // TODO Auto-generated method stub
    }
    @Override
    public void onLocationChanged(Location location) {
        // TODO Auto-generated method stub
        String string = "纬度为：" + location.getLatitude() + "\n 经度为：" + location.
                                                                      getLongitude();
        //Toast.makeText(getApplicationContext(), string2, Toast.LENGTH_LONG).show();
        tv.setText(string);    //刷新 UI 控件
    }
});
```

【例 10.2.1】 Android 三种定位方式的综合运用。

初始部署、运行时，依次进行是否有定位权限、是否开启了定位服务和是否有网络连接的检查。若没有定位权限，可动态申请（出现权限授予对话框）；若没有开启定位服务或者没有网络连接，则进入相应的设置程序。最后，以某种方式定位，并将定位结果刷新 UI 控件，程序流程如图 10.2.2 所示。

主要设计步骤如下：

（1）新建名为 Example10 的项目，再新建名为 examle10_1 的模块。
（2）在模块的清单文件里，注册定位权限，其代码如下：

```
<!-- 2 个定位权限是危险权限并处于一个权限组，需要动态申请 -->
<uses-permission android:name="android.permission.ACCESS_FINE_LOCATION"/><!--精确定位-->
<!--粗糙定位-->
<uses-permission android:name="android.permission.ACCESS_COARSE_LOCATION"/>
<!-- 网络权限和 WiFi 权限是普通权限，不需要动态申请 -->
<!--网络状态-->
<uses-permission android:name="android.permission.ACCESS_NETWORK_STATE"/>
<!--WiFi 状态-->
<uses-permission android:name="android.permission.ACCESS_WIFI_STATE"/>
```

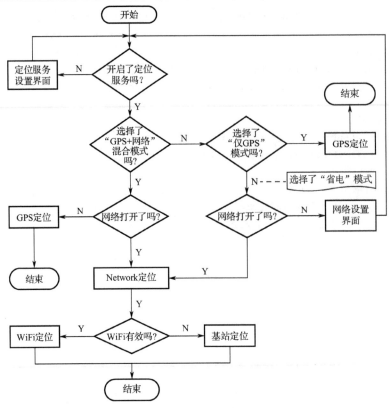

图 10.2.2 程序流程

（3）编写 Activity 组件 MainActivity，其代码如下：

```
/*
    打开位置服务时，默认选择混合定位
    GPS 定位精度高，但可能由于 GPS 信号弱而较慢甚至获取不到
    GPS 定位对应于权限 android.permission.ACCESS_FINE_LOCATION
    网络定位（WiFi 或移动网络）定位速度快，但精度相对 GPS 低
    网络定位对应于权限 android.permission.ACCESS_COARSE_LOCATION【COARSE 系粗糙之意】
*/
public class MainActivity extends AppCompatActivity {
    LocationManager lm;   //位置服务管理器
    ConnectivityManager cm;   //网络连接管理器
    WifiManager wfm;   //WiFi 管理器
    TextView tv;   //显示经度和纬度
    final static int LOCATION_SETTING_REQUEST_CODE = 100;
    final static int NETWORK_SETTING_REQUEST_CODE = 200;
    @Override
    protected void onCreate(Bundle savedInstanceState) {
        super.onCreate(savedInstanceState);
        setContentView(R.layout.activity_main);
        //如果没有 ACCESS_COARSE_LOCATION 权限，动态请求用户允许使用该权限
```

```java
        if (ActivityCompat.checkSelfPermission(this, Manifest.permission.ACCESS_COARSE_
                                        LOCATION)
                                != PackageManager.PERMISSION_GRANTED) {
            ActivityCompat.requestPermissions(this, new String[]
                            {Manifest.permission.ACCESS_COARSE_LOCATION}, 1);
        }else {
            prepareLocatingService();
        }
    }
    @Override
    public void onRequestPermissionsResult(int requestCode, @NonNull String[] permissions,
                                        @NonNull int[] grantResults) {
        super.onRequestPermissionsResult(requestCode, permissions, grantResults);
        switch (requestCode) {
            case 1:
                if (grantResults[0] == PackageManager.PERMISSION_GRANTED) {    //危险权限
                    prepareLocatingService();
                } else {
                    Toast.makeText(this, "没有授予定位权限！ ", Toast.LENGTH_LONG).
                                                                    show();
                    finish();
                }
        }
    }
    void prepareLocatingService() {   //定位服务预备
        tv = findViewById(R.id.tv);
        lm = (LocationManager) getSystemService(Context.LOCATION_SERVICE);
        cm = (ConnectivityManager) getSystemService(Context.CONNECTIVITY_SERVICE);
        //android API 24 之后，必须使用 getApplicationContext()
        wfm = (WifiManager) getApplicationContext().getSystemService(Context.WIFI_
                                                                    SERVICE);
        //是否打开定位服务，有多种检测方法
        if (lm.isProviderEnabled(LocationManager.GPS_PROVIDER) ||
                        lm.isProviderEnabled(LocationManager.NETWORK_PROVIDER)) {
            locatingService();   //定位的入口，方法调用
        } else {
            new AlertDialog.Builder(this)
                    .setIcon(R.mipmap.ic_launcher)
                    .setTitle("消息框")
                    .setMessage("请先打开定位服务！ ")
                    .setPositiveButton("确定", new DialogInterface.OnClickListener() {
                        @Override
                        public void onClick(DialogInterface dialog, int which) {
                            //定位服务设置意图
                            Intent intent = new Intent(
```

```java
                            Settings.ACTION_LOCATION_SOURCE_SETTINGS);
                    //有返回值的调用,设置定位设置请求码
                            startActivityForResult(intent, LOCATION_SETTING_REQUEST_
                                                                              CODE);
                        }
                    }).show();
                //未阻塞,即用户单击"确定"按钮前此信息已经输出
                Log.i("wustwzxabc", "test1111");
                //即 Android 里的 AlertDialog,是非阻塞的!
            }
        }
        void locatingService() {    //已开启位置服务时
            if (lm.isProviderEnabled(LocationManager.GPS_PROVIDER) &&
                            lm.isProviderEnabled(LocationManager.NETWORK_PROVIDER)) {
                if (cm.getActiveNetworkInfo() != null) {    //使用混合定位(GPS+网络)
                    networkLocation();    //为了尽快获取定位数据,假定网络定位优先
                } else {
                    GPSLocation();    //使用 GPS 定位
                }
            } else {
                //使用仅 GPS 定位方式
                if (lm.isProviderEnabled(LocationManager.GPS_PROVIDER)) {
                    GPSLocation();
                } else {
                    if (lm.isProviderEnabled(LocationManager.NETWORK_PROVIDER)) {
                        networkLocation();
                    }
                }
            }
        }
        void GPSLocation() {
            tv.setText("GPS 定位中...");
            Toast.makeText(getApplicationContext(), "GPS 定位", Toast.LENGTH_LONG).show();
            if (ActivityCompat.checkSelfPermission(this, Manifest.permission.
                                                    ACCESS_FINE_LOCATION) !=
                                        PackageManager.PERMISSION_GRANTED) {
                return;
            }
            lm.requestLocationUpdates(LocationManager.GPS_PROVIDER, 1000, 1, new
                                                                LocationListener() {
                @Override
                public void onStatusChanged(String arg0, int arg1, Bundle arg2) {
                    // TODO Auto-generated method stub
                }
                @Override
                public void onProviderEnabled(String arg0) {
```

```java
                // TODO Auto-generated method stub
            }
            @Override
            public void onProviderDisabled(String arg0) {
                // TODO Auto-generated method stub
            }
            @Override
            public void onLocationChanged(Location location) {
                // TODO Auto-generated method stub
                String string = "纬度为：" + location.getLatitude() + "\n 经度为：" + location.
                                                                            getLongitude();
                tv.setText(string);
            }
        });
    }
    void networkLocation() {
        if (cm.getActiveNetworkInfo() != null) { //有网络连接时
            if (wfm.isWifiEnabled()) { //WiFi 有效时
                Toast.makeText(getApplicationContext(), "WiFi 方式的网络定位",
                                                    Toast.LENGTH_LONG).show();
            } else {
                Toast.makeText(getApplicationContext(), "基站方式的网络定位",
                                                    Toast.LENGTH_LONG).show();
            }
        } else { //没有网络连接时
            new AlertDialog.Builder(this)
                    .setIcon(R.mipmap.ic_launcher)
                    .setTitle("消息框")
                    .setMessage("无网络连接！请先设置一种网络连接...")
                    .setPositiveButton("确定", new DialogInterface.OnClickListener() {
                        @Override
                        public void onClick(DialogInterface dialog, int which) {
                            //进入网络连接设置
                            Intent intent = new Intent(Settings.ACTION_SETTINGS);
                            //有返回值的调用，设置网络设置请求码
                            startActivityForResult(intent, NETWORK_SETTING_REQUEST_
                                                                        CODE);
                        }
                    })
                    .show();
    }
    //下面的请求位置（实时）更新，
    //Android 6.0 及以上版本需要在首次运行时动态加载上述权限
    if (ActivityCompat.checkSelfPermission(this, Manifest.permission.ACCESS_ COARSE_
                                                                        LOCATION)
                                            != PackageManager.PERMISSION_GRANTED) {
```

```
            }
            lm.requestLocationUpdates(LocationManager.NETWORK_PROVIDER, 1000, 1,
                                                    new LocationListener() {   //实时监听
                @Override
                public void onStatusChanged(String arg0, int arg1, Bundle arg2) {
                    // TODO Auto-generated method stub
                }
                @Override
                public void onProviderEnabled(String arg0) {
                    // TODO Auto-generated method stub
                }
                @Override
                public void onProviderDisabled(String arg0) {
                    // TODO Auto-generated method stub
                }
                @Override
                public void onLocationChanged(Location location) {
                    // TODO Auto-generated method stub
                    String string = "纬度为：" + location.getLatitude() + "\n 经度为：" +
                                                            location.getLongitude();
                    //Toast.makeText(getApplicationContext(), string2, Toast.LENGTH_LONG).
                                                                                  show();
                    tv.setText(string);
                }
            });
        }
        @Override   //有返回值调用的回调
        protected void onActivityResult(int requestCode, int resultCode, @Nullable Intent data) {
            super.onActivityResult(requestCode, resultCode, data);
            //自定义的进入定位设置和网络设置时的请求码
            if (requestCode == LOCATION_SETTING_REQUEST_CODE || requestCode ==
                                                NETWORK_SETTING_REQUEST_CODE) {
                prepareLocatingService();
            }
        }
    }
```

注意：Android 原生的定位程序在 Android 6.0 和 7.0 设备能成功运行，而在 Android 8.0 及以上不能成功运行，解决的办法是：安装本定位应用程序后，进入设置界面，搜索定位（Location），关闭本应用程序的定位权限后再打开。其中，模拟器的应用程序定位权限在选项 App-level permissions 里，且默认的定位方式是 GPS 定位。

10.3 百度定位及地图应用开发

10.3.1 百度定位应用开发基础

1．百度定位应用概述

Android 定位 SDK 自 v7.0 版本起，按照附加功能不同，向开发者提供了如下 4 种不同类型的定位开发包。

基础定位：开发包体积最小，但只包含基础定位能力（GPS/WiFi/基站）、基础位置描述能力；

离线定位：在基础定位能力上，提供离线定位能力，可在网络环境不佳时，进行精准定位；

室内定位：在基础定位能力上，提供室内高精度定位能力，精度可达1～3m；

全量定位：包含离线定位、室内高精度定位能力，同时提供更人性化的位置描述服务。

注意：这四类开发包互斥，一个应用中只需集成一种定位开发包即可。

2．百度定位应用开发包下载及使用

访问 http://lbsyun.baidu.com，依次选择：开发文档→Android 地图 SDK→产品下载→自定义下载，选择定位包后下载并解压。切换项目视图至 Project 视图，将.jar 文件复制到 libs 文件夹后，右击 jar 文件，选择"Add As Library"项；将包含.so 文件的几个文件夹复制到文件夹 src/main/jniLibs 里，如图 10.3.1 所示。

图 10.3.1　引入百度定位及百度地图所需要的文件

3．百度定位包中的主要 API

在百度定位包 com.baidu.location 里，存放了定位客户端类 LocationClient、客户端操作类 LocationClientOption、位置监听接口 BDLocationListener 和位置类 BDLocation，如图 10.3.2 所示。

```
v C  LocationClient                              v I  BDLocationListener
    m  LocationClient(Context)                      m  onReceiveLocation(BDLocation): void
    m  registerLocationListener(BDLocationListener): void   v C  BDLocation
    m  setLocOption(LocationClientOption): void        m  getLatitude(): double
    m  start(): void                                    m  getLongitude(): double
    m  stop(): void                                     m  getCountry(): String
v C  LocationClientOption                               m  getDistrict(): String
    v E  LocationMode                                   m  getStreet(): String
        m  LocationMode()                               m  getAddrStr(): String
        f  Hight_Accuracy: LocationMode                 m  getProvince(): String
        f  Battery_Saving: LocationMode                 m  getCity(): String
        f  Device_Sensors: LocationMode                 m  getSatelliteNumber(): int
    m  LocationClientOption()                           m  getLocType(): int
    m  setLocationMode(LocationMode): void              f  TypeNone: int = 0
    m  setIsNeedAddress(boolean): void                  f  TypeGpsLocation: int = 61
    m  setScanSpan(int): void                           f  TypeNetWorkLocation: int = 161
```

图 10.3.2 百度定位主要 API

位置客户端 LocationClient 是百度定位的主体，其构造以上下文对象作为参数，具有注册位置监听器、设定定位参数、发起定位和停止定位等方法。

百度位置类 BDLocation 封装了获取经度和纬度数据、地名信息和卫星数量等方法。

位置客户端操作类 LocationClientOption 的内部类 LocationMode 定义了 3 种定位模式。

10.3.2 注册百度开发者账号，申请位置应用的 Key

创建一个百度位置应用，需要先申请一个百度开发者账号。如果已有账号，则要先登录后才能创建应用。访问 http://developer.baidu.com，可完成注册或登录，如图 10.3.3 所示。

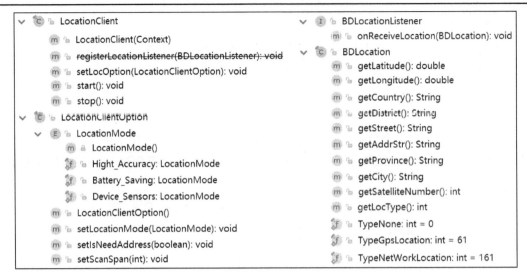

图 10.3.3 百度开发者登录界面

登录百度账号成功后，访问http://lbsyun.baidu.com/apiconsole/key，将进入应用的控制台。申请创建一个新的百度位置应用时，在输入"应用名称"、选择"应用类型"项后，

第10章 位置服务与地图应用开发

还需要输入本机的 Android 指纹码 SHA1 和应用的包名,如图 10.3.4 所示。

图 10.3.4　创建一个百度位置应用

提交后,将生成本应用的 Key,并显示在当前登录用户的控制台里。

在 Android Studio 里,获得 SHA1 的方法是单击屏幕右上方的 Gradle 工具后,展开项目的任何一个模块里的文件夹 Tasks/android,可以看到签名报告文件 SigningReport,如图 10.3.5 所示。

图 10.3.5　获取百度位置应用的签名信息

双击 signingReport,即可出现本机的 SHA1,如图 10.3.6 所示。

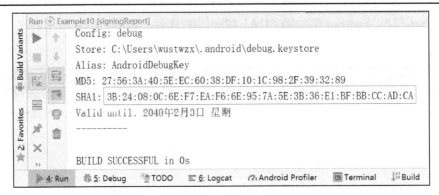

图 10.3.6 获取本机的 SHA1

注意：

（1）SHA（Secure Hash Algorithm），即安全散列算法，常用于数字签名。

（2）应用的包名是唯一的，它在清单文件里由属性 package 指定。

（3）每个 Key 仅且唯一对于一个 Android 应用验证有效。

10.3.3 在清单文件中注册权限、服务及应用的 Key

1．在清单文件中注册权限

百度定位应用开发，在清单文件里，需要注册的权限如下（共 6 个）：

```
<!--百度定位所需要权限，前面是 LOCATE 权限组的 2 个危险权限-->
<uses-permission android:name="android.permission.ACCESS_COARSE_LOCATION"/>
<uses-permission android:name="android.permission.ACCESS_FINE_LOCATION"/>
<!--百度定位所需要的普通权限-->
<uses-permission android:name="android.permission.ACCESS_WIFI_STATE"/>
<uses-permission android:name="android.permission.ACCESS_NETWORK_STATE"/>
<uses-permission android:name="android.permission.CHANGE_WIFI_STATE"/>
<!--因为程序要与百度云服务交互-->
<uses-permission android:name="android.permission.INTERNET"/>
```

2．在清单文件中注册百度云服务

由于每个百度定位 App 都拥有自己单独的定位 Service，因此，在使用百度定位及地图服务前，应在清单文件的<application>标签中声明 Service 服务组件，代码如下：

```
//需要权限 android.permission.INTERNET
<service
    android:name="com.baidu.location.f"
    android:enabled="true"
    android:process=":remote" />
```

3．在清单文件中注册应用的 Key

在创建了百度位置应用的 Key 后，需要在清单文件的<meta>标签内登记，示例代码

如下：

```
<meta-data
    android:name="com.baidu.lbsapi.API_KEY"
    android:value="NDnYRd8LK85vGbLeRVaV23x9RccY5M3p"/>
<!--应用 Key 是在百度开发者页面里生成的，需要替换-->
```

10.3.4 百度综合定位实现

以经纬度形式提供的定位结果是不直观的。百度定位 API，提供了获取地名等信息的相关方法。

【例 10.3.1】 百度综合定位并实时显示位置、地名等信息。

在室内程序运行时，先检测手机是否有网络连接，若没有，则运行手机设置程序。设置网络连接返回后，即可显示手机当前位置的经纬度、街道地名等信息，如图 10.3.6 所示。

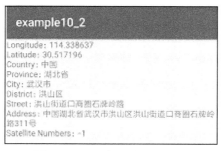

图 10.3.6 百度综合定位信息

主要设计步骤如下：

（1）在 Example10 项目里，再新建名为 examle10_2 的模块。

（2）访问 http://lbsyun.baidu.com，下载百度定位应用的软件包。解压后，分别将.jar 文件和.so 文件复制到模块的相关文件夹里并引用。

（3）访问 http://developer.baidu.com，注册百度开发者账号，创建本应用的 Key。

（4）在模块的清单文件里，注册定位所需的 6 个权限，注册百度云服务组件，注册已经生成的应用 Key。

（5）编写 Activity 组件 MainActivity，其代码如下：

```
/*
    百度定位信息，使用一个 TextView 控件显示
    一般需要打开定位服务，选择高精度定位模式，有网络连接
    需要在清单文件里使用百度云服务（参见清单文件 Service 标签）
    需要创建应用（模块）的 Key，并写入清单文件（参见清单文件 Meta 标签）
*/
public class MainActivity extends AppCompatActivity {
    LocationClient mLocationClient; //定位客户端
    TextView tv_positionText;    //显示定位信息控件
```

```java
@Override
protected void onCreate(Bundle savedInstanceState) {
    super.onCreate(savedInstanceState);
    setContentView(R.layout.activity_main);
    //如果没有定位权限，动态请求用户允许使用该权限
    if (ActivityCompat.checkSelfPermission(this, Manifest.permission.ACCESS_FINE_
                                                                    LOCATION) !=
                                    PackageManager.PERMISSION_GRANTED) {
        ActivityCompat.requestPermissions(this, new String[]
                              {Manifest.permission.ACCESS_FINE_LOCATION}, 1);
    } else {
        requestLocation();
    }
}
@Override
public void onRequestPermissionsResult(int requestCode, @NonNull String[] permissions,
                                       @NonNull int[] grantResults) {
    switch (requestCode) {
        case 1:
            if (grantResults[0] != PackageManager.PERMISSION_GRANTED) {
                Toast.makeText(this, "没有定位权限！", Toast.LENGTH_LONG). show();
                finish();
            } else {
                requestLocation();
            }
    }
}
private void requestLocation() {
    initLocation();    //初始化
    mLocationClient.start();    //开始定位
}
@Override
protected void onDestroy() {
    super.onDestroy();
    mLocationClient.stop();
}
private void initLocation() {
    //创建位置客户端
    mLocationClient = new LocationClient(getApplicationContext());
    //注册位置监听器
    mLocationClient.registerLocationListener(new MyLocationListener());
    tv_positionText = findViewById(R.id.tv_positionText);
    //定位客户端操作
    LocationClientOption option = new LocationClientOption();
    //设置扫描时间间隔（单位：ms）
    option.setScanSpan(1000);
```

```
            //设置定位模式，三选一
            option.setLocationMode(LocationClientOption.LocationMode.Hight_Accuracy);
            /*option.setLocationMode(LocationClientOption.LocationMode.Battery_Saving);
            option.setLocationMode(LocationClientOption.LocationMode.Device_Sensors);*/
            //设置需要地址信息
            option.setIsNeedAddress(true);
            //设置客户端定位参数
            mLocationClient.setLocOption(option);
        }
        class MyLocationListener implements BDLocationListener {
            @Override
            public void onReceiveLocation(BDLocation bdLocation) {
                StringBuffer currentPosition = new StringBuffer()
                    .append("Longitude：").append(bdLocation.getLongitude()).append("\n")
                    .append("Latitude：").append(bdLocation.getLatitude()).append("\n")
                    .append("Country：").append(bdLocation.getCountry()).append("\n")
                    .append("Province：").append(bdLocation.getProvince()).append("\n")
                    .append("City：").append(bdLocation.getCity()).append("\n")         //市
                    .append("District：").append(bdLocation.getDistrict()).append("\n")  //区
                    .append("Street：").append(bdLocation.getStreet()).append("\n")     //街道
                    .append("Address：").append(bdLocation.getAddrStr()).append("\n")   //完整地址
                    .append("Satellite Numbers：").append(bdLocation.getSatelliteNumber()); //卫星数
                tv_positionText.setText(currentPosition);    //刷新 UI 控件
            }
        }
    }
```

（6）部署项目到手机后，做运行测试。

10.3.5 百度地图显示

以地图形式显示当前位置信息，是更加直观的方式。显示一个地图，只需要在布局文件中添加一个 MapView 控件即可。但是，想要对其进行一些控制，就需要对 MapView 控件应用方法 getMap()，返回一个 BaiduMap 对象。

【例 10.3.2】 显示当前位置的地图。

程序运行时，先检测手机是否开启网络连接，若没有开启，则运行手机设置程序。返回后，逐渐出现定位信息和以当前位置为中心的地图，如图 10.3.7 所示。

主要设计步骤如下：

（1）在 Example10 项目里，再新建名为 examle10_3 的模块，引用百度定位包。

（2）访问 http://developer.baidu.com，登录百度开发者账号后，创建本应用的 Key。

（3）在模块的清单文件里，注册定位所需的 6 个权限，注册百度云服务组件，注册已经生成的应用 Key。

图 10.3.7 以地图形式显示当前位置

（4）布局文件使用 FrameLayout（帧）布局，使用百度提供的 MapView 控件，实现位置文本信息与地图的重叠效果。布局文件代码如下：

```
<FrameLayout xmlns:android="http://schemas.android.com/apk/res/android"
    android:layout_width="match_parent"
    android:layout_height="match_parent" >
    <!--百度地图控件-->
    <com.baidu.mapapi.map.MapView
        android:id="@+id/bmapView"
        android:layout_width="fill_parent"
        android:layout_height="fill_parent"
        android:clickable="true" />
    <!--位置文本信息与地图因为使用帧布局而有重叠效果,背景色前面的2位代码为透明度-->
    <LinearLayout
        android:layout_width="fill_parent"
        android:layout_height="wrap_content"
        android:background="#e0000000"
        android:orientation="vertical" >
        <LinearLayout
            android:layout_width="wrap_content"
            android:layout_height="wrap_content"
            android:layout_marginLeft="12dp"
            android:layout_marginTop="20dp"
```

```xml
            android:orientation="horizontal" >
            <TextView
                android:layout_width="wrap_content"
                android:layout_height="wrap_content"
                android:text="纬度："
                android:textColor="#ffffff"
                android:textSize="15dp" />
            <TextView
                android:id="@+id/tv_Lat"
                android:layout_width="wrap_content"
                android:layout_height="wrap_content"
                android:text=""
                android:textColor="#ffffff"
                android:textSize="15dp" />
        </LinearLayout>
        <LinearLayout
            android:layout_width="wrap_content"
            android:layout_height="wrap_content"
            android:layout_marginLeft="12dp"
            android:layout_marginTop="10dp"
            android:orientation="horizontal" >
            <TextView
                android:layout_width="wrap_content"
                android:layout_height="wrap_content"
                android:text="经度："
                android:textColor="#ffffff"
                android:textSize="15dp" />
            <TextView
                android:id="@+id/tv_Lon"
                android:layout_width="wrap_content"
                android:layout_height="wrap_content"
                android:text=""
                android:textColor="#ffffff"
                android:textSize="15dp" />
        </LinearLayout>
        <LinearLayout
            android:layout_width="wrap_content"
            android:layout_height="wrap_content"
            android:layout_marginBottom="10dp"
            android:layout_marginLeft="12dp"
            android:layout_marginTop="10dp"
            android:orientation="horizontal" >
            <TextView
                android:layout_width="wrap_content"
                android:layout_height="wrap_content"
                android:text="地址："
```

```xml
            android:textColor="#ffffff"
            android:textSize="15dp" />
        <TextView
            android:id="@+id/tv_Add"
            android:layout_width="wrap_content"
            android:layout_height="wrap_content"
            android:text=""
            android:textColor="#ffffff"
            android:textSize="15dp" />
    </LinearLayout>
  </LinearLayout>
</FrameLayout>
```

（5）编写 Activity 组件 MainActivity，其代码如下：

```java
/*
    百度地图应用，包含定位信息和地图显示
    一般需要打开定位服务，选择高精度定位模式，有网络连接
    需要在清单文件里使用百度云服务（参见清单文件 Service 标签）
    需要创建应用（模块）的 Key，并写入清单文件（参见清单文件 Meta 标签）
*/
public class MainActivity extends AppCompatActivity {
    LocationClient mLocationClient;  //定位客户端
    MapView mapView;    //Android Widget 地图控件
    BaiduMap baiduMap;   //封装了对地图对象的各种手势操作
    boolean isFirstLocate = true;
    TextView tv_Lat;    //纬度
    TextView tv_Lon;    //经度
    TextView tv_Add;    //地址
    @Override
    protected void onCreate(Bundle savedInstanceState) {
        super.onCreate(savedInstanceState);
        //如果没有定位权限，动态请求用户允许使用该权限
        if (ActivityCompat.checkSelfPermission(this, Manifest.permission.
        ACCESS_FINE_LOCATION) != PackageManager.PERMISSION_GRANTED) {
            ActivityCompat.requestPermissions(this, new String[]
                            {Manifest.permission.ACCESS_FINE_LOCATION}, 1);
        }else {
            requestLocation();
        }
    }
    @Override
    public void onRequestPermissionsResult(int requestCode, @NonNull String[] permissions,
                                            @NonNull int[] grantResults) {
        switch (requestCode) {
            case 1:
```

```java
            if (grantResults[0] != PackageManager.PERMISSION_GRANTED) {
                Toast.makeText(this, "没有定位权限！", Toast.LENGTH_LONG).show();
                finish();
            } else {
                requestLocation();
            }
        }
    }
    private void requestLocation() {
        //定位前初始化
        initLocation();
        //发起定位
        mLocationClient.start();
    }
    private void initLocation() {    //初始化
        mLocationClient = new LocationClient(getApplicationContext());
        mLocationClient.registerLocationListener(new MyLocationListener());
        //初始化地图应用
        SDKInitializer.initialize(getApplicationContext());
        setContentView(R.layout.activity_main);
        mapView = findViewById(R.id.bmapView);
        baiduMap = mapView.getMap();
        tv_Lat = findViewById(R.id.tv_Lat);
        tv_Lon = findViewById(R.id.tv_Lon);
        tv_Add = findViewById(R.id.tv_Add);
        //定位客户端操作
        LocationClientOption option = new LocationClientOption();
        //设置扫描时间间隔
        option.setScanSpan(1000);
        //设置定位模式，三选一
        option.setLocationMode(LocationClientOption.LocationMode.Hight_Accuracy);
        /*option.setLocationMode(LocationClientOption.LocationMode.Battery_Saving);
        option.setLocationMode(LocationClientOption.LocationMode.Device_Sensors);*/
        //设置需要地址信息
        option.setIsNeedAddress(true);
        //保存定位参数
        mLocationClient.setLocOption(option);
    }
    //内部类，百度位置监听器
    private class MyLocationListener    implements BDLocationListener {
        @Override
        public void onReceiveLocation(BDLocation bdLocation) {
            tv_Lat.setText(bdLocation.getLatitude()+"");
            tv_Lon.setText(bdLocation.getLongitude()+"");
            tv_Add.setText(bdLocation.getAddrStr());
            //GPS定位或网络定位时
```

```
                if(bdLocation.getLocType()==BDLocation.TypeGpsLocation ||
                        bdLocation.getLocType()==BDLocation.TypeNetWorkLocation){
            navigateTo(bdLocation);
        }
    }
}
private void navigateTo(BDLocation bdLocation) {
    if(isFirstLocate){
        LatLng ll = new LatLng(bdLocation.getLatitude(),bdLocation.getLongitude());
        MapStatusUpdate update = MapStatusUpdateFactory.newLatLng(ll);
        //以动画更新方式，实现对手势（如移动、缩放等操作）引起的地图状态更新
        baiduMap.animateMapStatus(update);
        isFirstLocate = false;
    }
}
@Override
protected void onResume() {
    super.onResume();
    mapView.onResume();
}
@Override
protected void onPause() {
    super.onPause();
    mapView.onResume();
}
@Override
protected void onDestroy() {
    super.onDestroy();
    mLocationClient.stop();
    mapView.onDestroy();
}
}
```

（6）部署项目到手机后，做运行测试。

习 题 10

一、判断题

1．GPS 定位比网络定位省电。
2．Android WiFi 定位应用，需要 android.permission.INTERNET 权限。
3．方法 requestLocationUpdates()参数包含定位提供者和位置监听器等对象。
4．百度定位应用，必须在清单文件里注册 android.permission.INTERNET 权限。
5．做百度定位应用开发，必须注册百度开发者并申请应用 Key。

二、选择题

1．为了进行 GPS 定位，手机至少需要接收到____颗卫星信号。
   A．2　　　　　B．3　　　　　C．4　　　　　D．5
2．类 LocationManager 定义在包____里。
   A．android.location　　　　　B．android.app
   C．android.content　　　　　D．android.media
3．下列选项中，不是 Android 合法类的是____。
   A．TelephonyManager　　　　　B．WifiManager
   C．GPSManager　　　　　D．ConectivityManager
4．使用 GPS 定位，必须开启的权限是____。
   A．GLOBAL_SEARCH
   B．ACCESS_FINE_LOCATION
   C．ACCESS_COARSE_LOCATION
   D．ACCESS_LOCATION_EXTRA_COMMANDS
5．下列类常量中，表示网络定位服务的是____。
   A．Context.LOCATION_SERVICE
   B．LocationManager.GPS_PROVIDER
   C．LocationManager.NETWORK_PROVIDER
   D．Context.WIFI_SERVICE

三、填空题

1．实现 Android 实时定位，需要使用 LocationManager 提供的____方法。
2．定位精确度最高的是____。
3．使用 WiFi 定位或 GPRS 定位，需要在清单文件里注册的权限是相同的，该权限名称是____。
4．地图显示控件名称是____。

实验 10

一、实验目的

(1) 掌握 Android GPS 定位的系统服务及相关类。
(2) 掌握 Android WiFi 定位的系统服务及相关类。
(3) 掌握百度定位相关 API 的使用。
(4) 掌握百度地图应用开发的基本方法。

二、实验内容及步骤

【预备】在 Android Studio 中，新建名为 Example10 的项目后，访问本课程配套的网站 http://www.wustwzx.com/as/sy/sy10.html，复制相关代码，完成如下几个模块的设计。

1. GPS 定位程序设计（参见例 10.2.1）

(1) 在项目 Example10 里，新建名为 example10_1 的模块，在其清单文件里，注册 Android 标准定位所需要的权限。

(2) 在 MainActivity 组件里，定义 LocationManager 类型的对象 lm、ConnectivityManager 类型的对象 cm 和 WifiManager 类型的对象 wfm 作为类成员。

(3) 在 MainActivity 的 onCreate()方法里，动态申请危险权限，并重写请求权限的回调方法 onRequestPermissionsResult()。

(4) 在自定义的定位方法里，分别实例化类的成员对象，编写使用 GPS 和网络实时定位的实现代码。

(5) 部署本模块，做运行测试。

2. 百度综合定位程序设计（参见例 10.3.1）

(1) 在项目 Example10 里，新建名为 example10_2 的模块，在其清单文件里，注册百度定位所需要的权限。

(2) 访问 http://developer.baidu.com，注册百度开发账号后登录，创建本模块的应用 Key。然后，在清单文件里，注册百度位置服务的百度云服务，注册应用的 Key。

(3) 在 MainActivity 的 onCreate()方法里，动态申请和处理定位的危险权限。

(4) 在 MainActivity 里，定义类成员 LocationClient mLocationClient，编写接口 BDLocationListener 的实现类作为 MainActivity 类的内部类。

(5) 在 MainActivity 里，编写定位实现代码，主要包含 mLocationClient 的实例化及其定位参数设置、位置监听器设置和发动定位等。

(6) 部署本模块，做运行测试。

3. 百度地图应用程序设计（参见例 10.3.2）

（1）在项目 Example10 里，新建名为 example10_3 的模块，在其清单文件里，注册百度定位所需要的权限（与模块 example10_2 相同）。

（2）访问 http://developer.baidu.com，注册百度开发账号后登录，创建本模块的应用 Key。然后，在清单文件里，注册百度定位服务的百度云服务，注册其应用的 Key。

（3）布局文件使用 FrameLayout（帧）布局，使用百度提供的 MapView 控件，实现位置文本信息与地图的重叠效果。

（4）在 MainActivity 的 onCreate()方法里，动态申请和处理定位的危险权限。

（5）在 MainActivity 里，分别定义类型 LocationClient 和 BaiduMap 的成员，编写接口 BDLocationListener 的实现类作为 MainActivity 类的内部类。

（6）在 MainActivity 里，编写定位及地图显示的实现代码，主要包含 LocationClient 对象及 BaiduMap 对象的实例化、设置定位参数、位置监听器设置、地图对手势的响应方式和发动定位等。

（7）部署本模块，做运行测试。

三、实验小结及思考

(由学生填写，重点写上机中遇到的问题。)

第 11 章 Android 网络编程

随着 Android 平台市场占有率的稳步上升，Android 应用的数量和种类越来越多，涉及的范围也越来越大。如大量的手机 App（如支付宝和美团等）被广泛地应用。开发者希望将手机变成互联网的移动终端，以扩展互联网应用的广度和深度；企业希望在手机平台上实现更多的管理和应用，系统能随时随地保持沟通，进而使企业低成本、高效率地运营。这些需求更多地表现在 Android 的互联网应用方面，其技术核心正是 Android 网络编程的相关知识，本章学习要点如下：

- 掌握 HttpURLConnection 访问网络资源的编程使用方法；
- 掌握网络接口 HttpClient 调用 Web 服务的编程方法；
- 掌握 Android 网络图片下载框架 Glide 的使用方法；
- 掌握 Android 网络编程框架 Volley 的使用方法；
- 掌握手机客户端与 Web 服务器通信的编程方法。

11.1 基于 HTTP 协议的 Android 网络编程

11.1.1 Android 网络编程概述

Android API 除了包含 Java 提供的标准网络编程方法外，还引入了 Apache 的 HTTP 扩展包，并针对 WiFi、Bluetooth 等设备分别提供了单独的 API。因此，在 Android 平台中，可以使用 Java 标准接口、Apache 接口和 Android 网络接口。

HTTP 是 HyperText Transfer Protocol 的英文缩写，表示超文本传输协议。HTTP 协议是互联网上应用最多、最为广泛的一种网络协议，是从 WWW 服务器传输超文本到本地浏览器的传送协议。

Android 网络编程可以划分为对应用层的 HTTP 编程和对传输层的 Socket 编程。

基于 HTTP 网络编程，既可以使用标准 Java 接口（对应于软件包 java.net），也可以使用 Apache 接口（对应于软件包 org.apache.http.client 和 org.apache.http.impl）或者 Android 网络接口（对应于软件包 android.net 和 android.net.http）。

由于对于网络状况的不可预见性，很有可能在网络访问时造成阻塞 UI 主线程（出现假死）的现象。解决该问题有如下方法：

- 独立线程；
- 异步线程 AsyncTask；
- StrictMode 修改默认的策略。

基于 HTTP 请求的 Android 应用，都需要在清单文件中配置如下访问 Internet 的权限：

<uses-permission android:name="android.permission.INTERNET"/>

注意：

（1）不能在 UI 主线程中访问网络，否则会阻塞 UI 操作。

（2）不能在子线程中更新 UI，应使用异步请求，如 Handler 机制。

11.1.2　HTTP 请求与响应

HTTP 有 Get 和 Post 两种请求方式。

Get 请求方式是通过把参数键值对附加在 URL 后面来传递的。在服务器端可以直接读取，效率较高，但缺乏安全性，也无法处理复杂的数据，长度受限制。Get 请求方式主要用于传递简单的参数。

在 Post 请求方式中，传输参数会被打包在 HTTP 报头中，可以是二进制的。Post 请求方式便于传送较大的数据，同时因为不会暴露数据在浏览器的地址栏中，所以安全性相对较高，但处理效率会受到影响。

11.1.3　使用 HttpURLConnection 访问网络资源

获取请求指定 URL 后响应信息的一种方法，它使用终结类 java.net.URL 的 openConnection()方法得到一个 URLConnection 对象，通过该对象建立与服务器的连接，进而获取服务器响应的信息，最后通过 java.io 包提供的相关类进行输出。

终结类 URL 与抽象类 URLConnection 的定义，如图 11.1.1 所示。

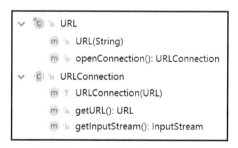

图 11.1.1　终结类 URL 与抽象类 URLConnection 的定义

通过 URL 类的 openConnection()方法创建 HTTP 请求的连接对象时，为了使用断开连接等方法，需要进行类型转换，其代码如下：

HttpURLConnection urlConn = (HttpURLConnection) url.openConnection();

注意：

（1）抽象类 HttpURLConnection 是抽象类 URLConnection 的子类，提供了断开 HTTP

连接的方法 disconnect()，父类不具有该方法。

（2）获取请求指定的 URL 后使用 org.apache.http.client. HttpClient 接口。

【例 11.1.1】 通过创建和使用 URL 对象，访问网络资源（获取网页源代码）。

程序运行时，单击界面中的"点击获取数据"按钮后，会在一个 TextView 控件里显示访问网站 http://www.wustwzx.com 主页时获取的 HTML 代码（含 JavaScript 脚本），其效果如图 11.1.2 所示。

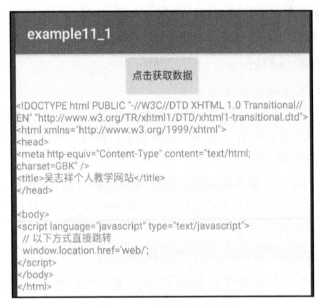

图 11.1.2　模块运行效果

由于 Android 网络通信，需要在非主线程中进行。因此，在 MainActivity 里创建一个内部类，其结构如图 11.1.3 所示。

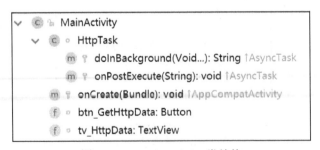

图 11.1.3　MainActivity 类结构

程序文件 MainActivity.java，其代码如下：

```
//显示访问网站页面的源代码
public class MainActivity extends AppCompatActivity {
    Button btn_GetHttpData;
    TextView tv_HttpData;
```

```java
@Override
protected void onCreate(Bundle savedInstanceState) {
    super.onCreate(savedInstanceState);
    setContentView(R.layout.activity_main);
    btn_GetHttpData = findViewById(R.id.btn_GetHttpData);
    tv_HttpData = findViewById(R.id.tv_HttpData);
    btn_GetHttpData.setOnClickListener(new View.OnClickListener() {
        @Override
        public void onClick(View v) {     //事件处理
            new HttpTask().execute();
        }
    });
}
class HttpTask extends AsyncTask<Void,Void,String>{
    @Override
    protected String doInBackground(Void... voids) {   //耗时操作代码在后台进行
        String httpUrl = "http://www.wustwzx.com"; //域名字符串
        String resultData = "";    //结果字符串
        URL url = null;   //URL 对象
        try {
            url = new URL(httpUrl);   //构造 URL 对象时需要使用异常处理
        } catch (MalformedURLException e) {
            Log.d("TAG", "URL 对象创建失败！");
        }
        if(url != null) {    //如果 URL 不为空时
            try {
                //有关网络操作时，需要使用异常处理
                HttpURLConnection urlConn = (HttpURLConnection) url.
                                                            openConnection();
                //服务器返回数字符流，网页文档编码一般为 UTF-8 或 GBK，据实选用
                InputStreamReader in = new InputStreamReader(urlConn.getInputStream(),
                                                            "GBK");
                //为输出创建 BufferedReader
                BufferedReader buffer = new BufferedReader(in);
                String inputLine = null;
                while (((inputLine = buffer.readLine()) != null)) {
                    resultData += inputLine + "\n";    //换行
                }
                in.close();    //关闭输入流
                urlConn.disconnect();    //关闭 HTTP 连接
            } catch (IOException e) {
                resultData=e.getMessage();
            }
        } else {
            resultData = "url is null"; //当 url 为空时输出
        }
```

```
            return resultData;
    }
    @Override
    protected void onPostExecute(String resultData) {    //在后台数据提交后更新 UI 主线程
        if (resultData != null)
            tv_HttpData.setText(resultData);    //原生用法,更新 UI 工作在 onPostExecute()
                                                                      方法里
        else
            tv_HttpData.setText("Sorry,the content is null");
    }
 }
}
```

11.1.4 使用网络接口 HttpClient 调用 Web 服务

Web 服务是一个平台独立、松耦合、基于可编程的应用程序,可使用开放的 XML 标准来描述、发布、发现、协调和配置这些应用程序,用于开发分布式的互操作的应用程序。简单地说,Web 服务是远程的某个服务器对外公开的某种功能或方法,通过调用该服务以获得用户需要的信息。

Web 服务及其使用方式,如图 11.1.4 所示。

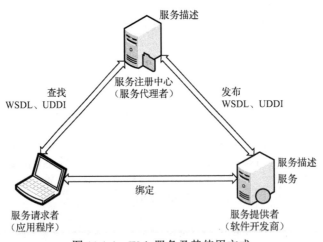

图 11.1.4　Web 服务及其使用方式

【例 11.1.2】 使用接口 HttpClient 调用 Web 服务,查询手机归属地。

程序运行后,输入手机号码(段)前面的 7 位号段,单击"查询"按钮后的程序运行界面,如图 11.1.5 所示。

从 Android API 23 起,Google 已经移除了 Apache HttpClient。在 Android Studio 中,为了使用 HttpClient 框架,需要对项目添加依赖库。使用项目结构工具,在管理模块依赖的搜索框(参见图 11.2.1)内输入"HttpClient"进行搜索,选择 cz.msebera.android:httpclient 即可。

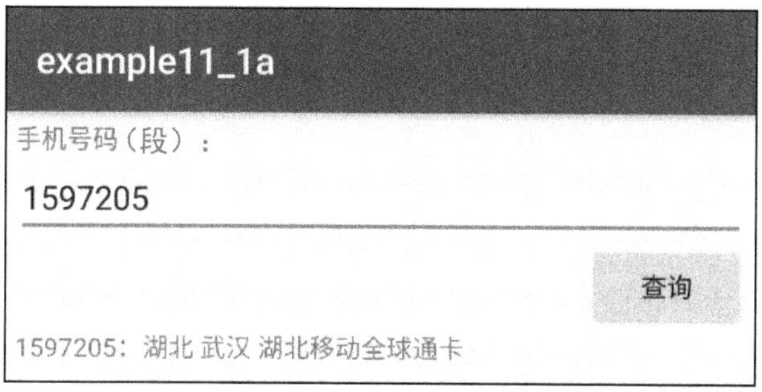

第 11 章　Android 网络编程

图 11.1.5　程序运行界面

程序文件 MainActivity.java，其代码如下：

```java
/*
    作为使用 Web 服务的示例程序，查询手机归属地
    由于对于网络状况的不可预见性，很有可能阻塞 UI 主线程出现假死的现象
    解决该问题的办法有：1．独立线程；2．异步线程 AsyncTask；3．StrictMode 修改默认的策略
    本项目只需要 Internet 网络权限，无危险权限！
*/
public class MainActivity extends AppCompatActivity {
    EditText phoneSecEditText;    //手机号段输入文本框
    TextView resultView ;    //结果标签
    Button query_btn;    //按钮
    @Override
    protected void onCreate(Bundle savedInstanceState) {
        super.onCreate(savedInstanceState);
        setContentView(R.layout.activity_main);
        //在主线程中强制使用子线程
        StrictMode.ThreadPolicy policy = new StrictMode.ThreadPolicy.Builder(). permitAll().
                                                                                build();
        StrictMode.setThreadPolicy(policy);
        phoneSecEditText = findViewById(R.id.phone_sec);
        resultView = findViewById(R.id.result_text);
        Button queryButton = findViewById(R.id.query_btn);
        queryButton.setOnClickListener(new View.OnClickListener() {
            @Override
            public void onClick(View v) {
                String phoneSec = phoneSecEditText.getText().toString().trim();    //手机号码（段）
                // 简单判断用户输入的手机号码（段）是否合法
                if ("".equals(phoneSec) || phoneSec.length() < 7) {
                    phoneSecEditText.setError("您输入的手机号码（段）有误！ ");
                    phoneSecEditText.requestFocus();
                    resultView.setText("");    // 清空
                    return;
                }
                Log.i("TAG", "onClick: ");
```

```java
            getRemoteInfo(phoneSec);    //查询手机号码（段）信息
        }
    });
}
//手机号段归属地查询
public void getRemoteInfo(String phoneSec) {
    String requestUrl = "http://ws.webxml.com.cn/WebServices/MobileCodeWS.asmx/getMobileCodeInfo";
    HttpClient client = new DefaultHttpClient();    //创建接口 HttpClient 的实例
    HttpPost post = new HttpPost(requestUrl);    //创建 HttpPost 请求对象
    List<NameValuePair> params = new ArrayList<NameValuePair>();
    params.add(new BasicNameValuePair("mobileCode", phoneSec));    //设置需要传递的参数
    params.add(new BasicNameValuePair("userId", ""));
    try {
        post.setEntity(new UrlEncodedFormEntity(params,"utf-8"));    //设置 URL 编码
        HttpResponse response = client.execute(post);    //发送请求、获取响应对象
        String result="";
        //判断请求是否成功处理
        if (response.getStatusLine().getStatusCode() == HttpStatus.SC_OK) {
            result = EntityUtils.toString(response.getEntity(),"utf-8");
        } else {
            result ="没有查询结果";
        }
        //resultView.setText(result);    //内容为 XML 格式
        resultView.setText(filterHtml(result));    //使用正则式过滤标签
    } catch (Exception e) {
        e.printStackTrace();
        Toast.makeText(getBaseContext(), "出错了!", Toast.LENGTH_SHORT).show();
    }
}
//使用正则表达式过滤 XML 文档中的标签（如<、>及空格等）
private String filterHtml(String source) {
    if(null == source){
        return "";
    }
    return source.replaceAll("\\?['-]+>","").trim();
}
```

11.2 Android 网络图像下载与通信框架

11.2.1 网络图像下载框架 Glide

Glide 是一个面向 Android 快速和高效的开源媒体管理和图像异步加载的框架，支持

视频、图片和 GIF 动画 3 种类型的资源，对其进行获取、解码和显示操作。

Glide 拥有灵活的 API，允许开发人员自定义添加网络堆栈。Glide 默认使用 HttpUrlConnection 的网络堆栈，也可以使用 Google 的 Volley 库和 Squareas 的 OkHttp 库来替代。

在 Android 项目里，加载 Glide 框架的操作，如图 11.2.1 所示。

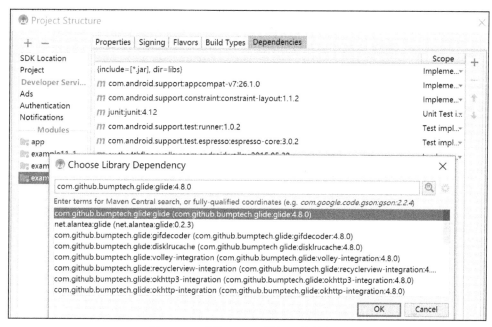

图 11.2.1　添加 Glide 框架（依赖包）

注意：Glide 框架 4.8.0 版本，会引入 com.android.support:appcompat-v7:27.0.2，可能与 build.gradle 的依赖包版本冲突。如果选用 Android API 版本为 26，则应将本模块 Gradle 脚本中版本号修改为 3.7.0 后重新下载。

在 Activity 程序里，使用 Glide 加载网络图像资源的示例代码如下：

```
ImageView pic = findViewById(R.id.imageView);
Glide.with(this)
        .load("http://www.wustwzx.com/webfront/projects/ShowPicture1/images/p1.jpg")
        .error(R.drawable.tb02)    //任选，加载网络图像失败时使用本地图像
        .into(pic);   //刷新图像控件
```

注意：使用 Glide 框架后，访问网络图片资源的代码不必放在子线程里。

11.2.2　网络通信框架 Volley

Google I/O 2013 发布了 Volley 框架，其具有如下的便利功能：
- JSON、图像等的异步下载；
- 网络请求的排序（scheduling）；
- 网络请求的优先级处理；

- 缓存；
- 多级别取消请求；
- 与 Activity 和生命周期的联动（Activity 结束时同时取消所有网络请求）。

在 Android Studio 项目里，加载 Volley 框架的操作，如图 11.2.2 所示。

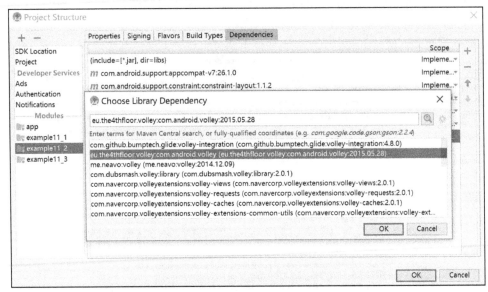

图 11.2.2　Volley 框架的引用

使用 Volley 框架，分为如下两个主要步骤。

（1）创建请求队列对象 RequestQueue，以缓存所有的 HTTP 请求。

Volley 类提供了创建 RequestQueue 的静态方法 newRequestQueue()，它以 Context 对象作为参数，如图 11.2.3 所示。

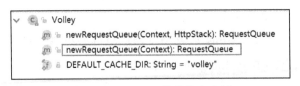

图 11.2.3　Volley 类的静态方法

（2）创建一个请求对象。

由于响应的结果类型主要有字符串、JSON 和图像 3 种类型，因此，Volley 框架提供了与之对应的类 StringRequest、JsonRequest 和 ImageRequest。其中，StringRequest 和 JsonRequest 只是返回的数据格式不同，前者是字符串，后者是 JSON 对象，使用方式类似。此外，JsonRequest 是抽象类，在实际开发中，应使用 JsonRequest 的子类 JsonObjectRequest 和 JsonArrayRequest。

（3）将请求对象添加至请求队列。

Volley 框架提供的请求类，如图 11.2.4 所示。

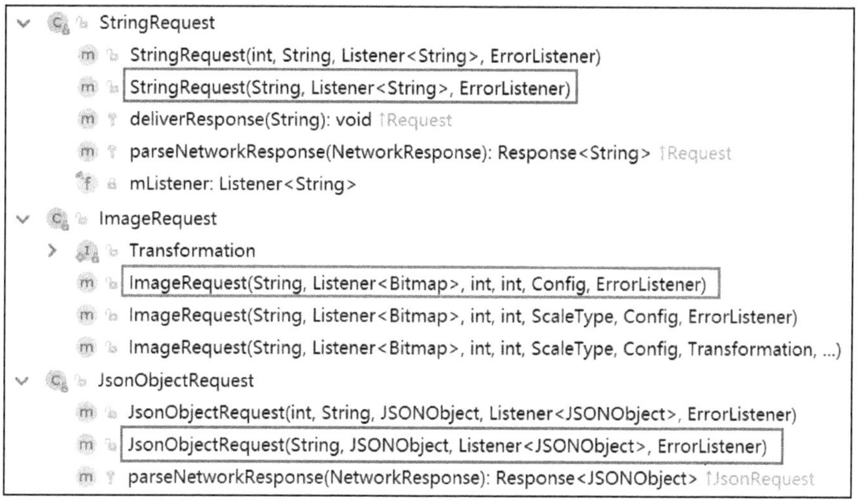

第 11 章　Android 网络编程

```
StringRequest
    StringRequest(int, String, Listener<String>, ErrorListener)
    StringRequest(String, Listener<String>, ErrorListener)
    deliverResponse(String): void ↑Request
    parseNetworkResponse(NetworkResponse): Response<String> ↑Request
    mListener: Listener<String>
ImageRequest
    Transformation
    ImageRequest(String, Listener<Bitmap>, int, int, Config, ErrorListener)
    ImageRequest(String, Listener<Bitmap>, int, int, ScaleType, Config, ErrorListener)
    ImageRequest(String, Listener<Bitmap>, int, int, ScaleType, Config, Transformation, ...)
JsonObjectRequest
    JsonObjectRequest(int, String, JSONObject, Listener<JSONObject>, ErrorListener)
    JsonObjectRequest(String, JSONObject, Listener<JSONObject>, ErrorListener)
    parseNetworkResponse(NetworkResponse): Response<JSONObject> ↑JsonRequest
```

图 11.2.4　Volley 请求类

Volley 框架的响应类 Response 包含两个内部接口 Listener 和 ErrorListener，分别负责响应结果和响应错误的回调，如图 11.2.5 所示。

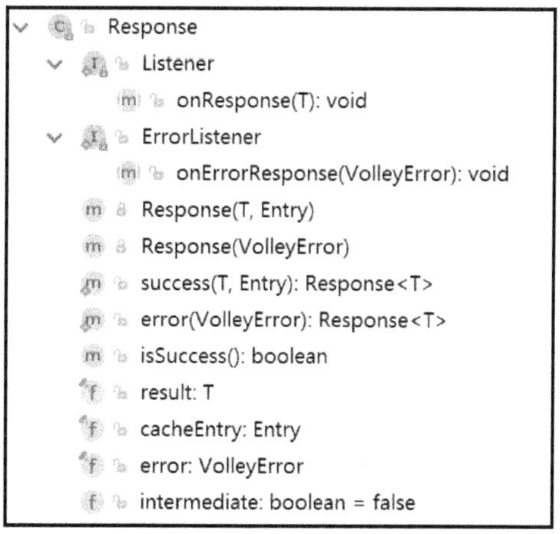

```
Response
    Listener
        onResponse(T): void
    ErrorListener
        onErrorResponse(VolleyError): void
    Response(T, Entry)
    Response(VolleyError)
    success(T, Entry): Response<T>
    error(VolleyError): Response<T>
    isSuccess(): boolean
    result: T
    cacheEntry: Entry
    error: VolleyError
    intermediate: boolean = false
```

图 11.2.5　Volley 响应类

在 Activity 程序里，使用 Volley 获取网络（图像）资源的示例代码如下：

```java
RequestQueue mRequestQueue= Volley.newRequestQueue(this);   //创建请求队列对象
StringRequest stringRequest=new StringRequest(   //文本请求
        "http://www.wustwzx.com",
        new Response.Listener<String>() {
            @Override
            public void onResponse(String response) {
                Log.i("TAG", response);
                Toast.makeText(MainActivity.this, response, Toast.LENGTH_LONG). show();
            }
        },
```

```java
            new Response.ErrorListener() {
                @Override
                public void onErrorResponse(VolleyError error) {
                    Log.e("TAG", error.getMessage(),error );
                }
            }
    );
    mRequestQueue.add(stringRequest);
    ImageRequest imageRequest = new ImageRequest(
            //构造方法包含 6 个参数
            "http://pic32.nipic.com/20130829/12906030_124355855000_2.png",
            //"http://www.wustwzx.com/webfront/projects/ShowPicture1/images/p1.jpg",
            new Response.Listener<Bitmap>() {
                @Override
                public void onResponse(Bitmap response) {
                    //要求定义类成员 ImageView pic;
                    pic.setImageBitmap(response);
                }
            },
            440, 320, Bitmap.Config.RGB_565,
            new Response.ErrorListener() {
                @Override
                public void onErrorResponse(VolleyError error) {
                    Toast.makeText(MainActivity.this, "未成功加载图片",
                            Toast.LENGTH_SHORT).show();
                    Log.e("TAG", error.getMessage(),error );
                }
            });
    mRequestQueue.add(imageRequest);
```

【例 11.2.1】 使用网络通信框架 Volley 获取网络资源。

模块运行效果，如图 11.2.6 所示。

【设计思路】选用 Volley 框架提供的请求类 JsonObjectRequest，获取 HTTP 请求返回的 JSON 对象形式的结果数据。对于普通文本，直接使用；对于图像的 URL，还要创建 ImageRequest 请求类来获取网络图像资源。

MainActivity 代码如下：

```java
/*
    Volley 框架的引入，可从项目结构的依赖里查看
    从豆瓣网 https://api.douban.com/v2/movie/subject/10606004 返回 JSON 数据，
    通过 JSON 在线格式化工具，可查看到包含的文本和图像信息
    访问 https://movie.douban.com，单击某个电影，查看地址栏里的 ID
    Volley 框架，简化了异步任务类的使用，UI 更新的原生用法，参见 example11_1
    扩展了图像下载功能，具有缓存 http 请求之功能
*/
public class MainActivity extends AppCompatActivity {
```

```java
Button btn_search;
EditText et_movieID;
TextView tv_title;
TextView tv_summary;
ImageView iv_stagePhoto;
String movieID;
String title;
String summary;
String stagePhotoUrl;
@Override
protected void onCreate(Bundle savedInstanceState) {
    super.onCreate(savedInstanceState);
    setContentView(R.layout.activity_main);
    btn_search = findViewById(R.id.btn_search);   //搜索按钮
    et_movieID = findViewById(R.id.et_movieID);   //影片编号
    tv_title = findViewById(R.id.tv_title);   //片名
    tv_summary = findViewById(R.id.tv_summary);   //剧情介绍
    iv_stagePhoto = findViewById(R.id.iv_stagePhoto);   //剧照
    btn_search.setOnClickListener(new View.OnClickListener() {
        @Override
        public void onClick(View v) {
            if ("".equals(et_movieID.getText().toString())) {
                movieID = "10606004?apikey=0b2bdeda43b5688921839c8ecb20399b";
            } else {
                movieID = et_movieID.getText().toString()+"?apikey=0b2bdeda43b56889218
                    39c8ecb20399b";
            }
            new HttpTask().execute(movieID);   //执行异步任务并输入一个 String 参数
        }
    });
}
class HttpTask extends AsyncTask<String, Void, Void> {   //泛型参数 Void 不能写成 void
    @Override
    protected Void doInBackground(String... strings) {   //Void 为 void 的包装类
        String baseUrl = "https://api.douban.com/v2/movie/subject/";
        String movieID = strings[0]; //调用时传递的参数
        String url = baseUrl + movieID;
        //使用 Volley 框架,先要创建请求队列
        final RequestQueue requestQueue = Volley.newRequestQueue(MainActivity.this);
        //JSON 对象请求,共 4 个参数
        JsonObjectRequest jsonObjectRequest = new JsonObjectRequest(
                url, //第 1 个参数
                null,   //第 2 个参数
                new Response.Listener<JSONObject>() {   //关键的第 3 个参数,请求成功
                                                       时的回调处理
                    @Override
                    public void onResponse(JSONObject response) {
```

```java
                    try {
                        title = response.getString("title");    //文本直接根据JSON对象
                                                                                                获取
                        tv_title.setText(title);
                        summary = response.getString("summary");
                        tv_summary.setText(summary);
                        stagePhotoUrl = response.getJSONObject("images"). getString
                                                                                    ("small");
                        ImageRequest imageRequest = new ImageRequest(
                                stagePhotoUrl,
                                new Response.Listener<Bitmap>() {
                                    @Override
                                    public void onResponse(Bitmap response) {
                                        iv_stagePhoto.setImageBitmap(response);
                                        Log.i("test", response.toString());
                                    }
                                }, 0, 0, Bitmap.Config.RGB_565,
                                new Response.ErrorListener() {
                                    @Override
                                    public void onErrorResponse(VolleyError error) {
                                        Toast.makeText(MainActivity.this, "未成功
                                                                                加载
                                        图片", Toast.LENGTH_SHORT).show();
                                    }
                                });
                        requestQueue.add(imageRequest);    //添加请求到队列
                    } catch (JSONException e) {
                        e.printStackTrace();
                    }
                }
            },
            new Response.ErrorListener() {    //第4个参数,请求失败时的回调处理
                @Override
                public void onErrorResponse(VolleyError error) {
                    Log.i("test", error.getMessage(), error);
                }
            });
        requestQueue.add(jsonObjectRequest);    //添加至队列
        return null;    //与前面doInBackground()方法返回值类型相适应
    }
  }
}
```

第 11 章　Android 网络编程

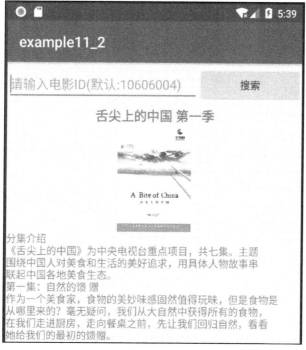

图 11.2.6　模块运行效果

11.3　手机 App 与 Web 服务器通信

11.3.1　Web 服务器项目

网站的 Web 服务器端有多种选择，在 Eclipse 中，常用 Tomcat 作为网站服务器、以 MySQL 作为数据库服务器。基于 Servlet 组件开发的用户登录项目结构，如图 11.3.1 所示。

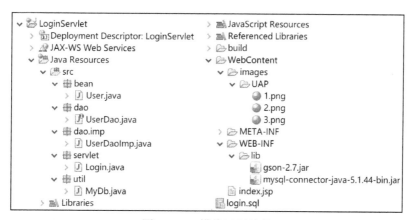

图 11.3.1　模块运行效果

Login.java 程序用于处理用户登录，其代码如下：

```java
/*
 * 需要导入包 com.google.gson.Gson
 * 输入用户名及密码
 * 返回登录结果的 JSON 字符串
 */
@WebServlet("/Login")
public class Login extends HttpServlet {
    private static final long serialVersionUID = 1L;
    protected void doPost(HttpServletRequest request, HttpServletResponse response)
            throws ServletException, IOException {
        response.setContentType("text/html");
        //接收用户输入
        String username = request.getParameter("username");
        String password = request.getParameter("password");
        //使用 DAO 层
        UserDao userDao = new UserDaoImp();
        int status;
        User user=null;
        List<User> friends=null;
        Map<String, Object> resultMap=new HashMap<String, Object>();      //登录结果
        if (userDao.isUserExsit(username)) { //用户名存在
            if ((user = userDao.query(username, password)) != null) {
                status = 1;     //登录成功
                friends = userDao.queryFriends(user.getUserID());
            } else {
                status = -2;    //密码错误
            }
        } else {
            status = -1;    //用户名不存在
        }
        resultMap.put("status", status);
        if (user != null) {
            resultMap.put("user", user);
        }
        if (friends != null) {
            resultMap.put("friends", friends);
        }
        PrintWriter out = response.getWriter();
        //使用 com.google.gson.Gson 转 Java Map 对象为 JSON 字符串
        out.print(new Gson().toJson(resultMap));
    }
    protected void doGet(HttpServletRequest request, HttpServletResponse response)
            throws ServletException, IOException {
        doPost(request, response);
    }
}
```

第 11 章　Android 网络编程

访问 http://localhost:8080/LoginServlet，表单输入（user1,1）后，输入返回 JSON 数据，如图 11.3.2 所示。

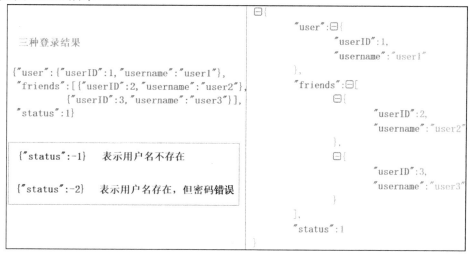

图 11.3.2　访问 JSP 网站时的登录结果

其中，图 11.3.2 右边是格式化 JSON 字符串的结果。

注意：JSON（JavaScript object notation）是一种独立于语言、轻量级的数据交换格式，采用键值对形式存储数据，常用 JSON 格式数据作为 API 的返回类型。

11.3.2　App 的登录程序设计

在没有固定 IP 的 Web 服务器情况下，为了能使用手机 App 与 Web 服务器通信，即手机 App 能访问 Web 服务器，需要自行搭建的 Web 服务器 IP 与手机 IP 在同一网段内，有如下 2 种方法在：

● 室内等局域网环境，只要手机和作为 Web 服务器的计算机都使用 WiFi 上网；
● 开启手机的 GPRS 网络及个人热点，使笔记本电脑通过手机 WiFi 上网。

在 Android 客户端，用户登录需要输入用户和密码，登录成功后进入好友列表界面，效果如图 11.3.3 所示。

图 11.3.3　用户登录 App

用户登录程序 LoginActivity.java 的类结构，如图 11.3.4 所示。

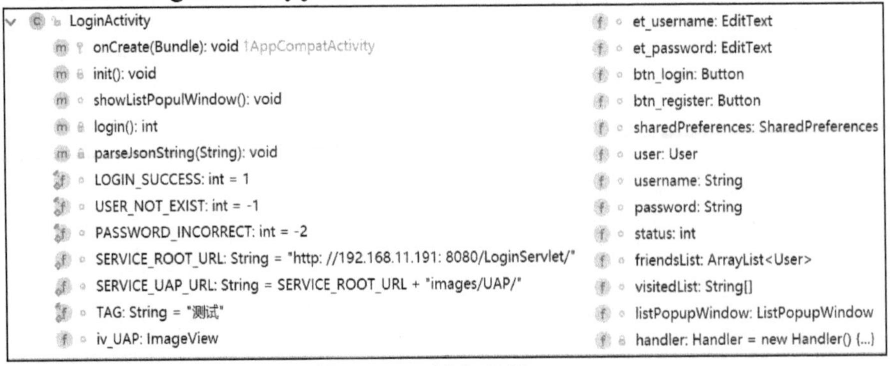

```
LoginActivity
    onCreate(Bundle): void  ↑AppCompatActivity      et_username: EditText
    init(): void                                    et_password: EditText
    showListPopulWindow(): void                     btn_login: Button
    login(): int                                    btn_register: Button
    parseJsonString(String): void                   sharedPreferences: SharedPreferences
    LOGIN_SUCCESS: int = 1                          user: User
    USER_NOT_EXIST: int = -1                        username: String
    PASSWORD_INCORRECT: int = -2                    password: String
    SERVICE_ROOT_URL: String = "http://192.168.11.191:8080/LoginServlet/"   status: int
    SERVICE_UAP_URL: String = SERVICE_ROOT_URL + "images/UAP/"   friendsList: ArrayList<User>
    TAG: String = "测试"                            visitedList: String[]
    iv_UAP: ImageView                               listPopupWindow: ListPopupWindow
                                                    handler: Handler = new Handler() {...}
```

图 11.3.4 用户登录 App

用户登录程序 LoginActivity.java 的代码如下：

```java
/*
本模块模仿 QQ 登录，需要访问 Web 服务器。运行本程序前，要求手机能访问 Web 服务器
为使 Web 服务器与手机在同一网段内，可以让计算机使用手机热点上网
在命令行方式下，Web 服务器 IP 使用 ipconfig/all 查看
LoginActivity 是界面程序，用于用户登录，其用户名及密码来源于数据库
本模块同时使用了网络编程框架 Volley 2015.05.28 和 Glide 3.7.0
登录成功后，将进入好友页面 HomeAtivity
注册功能，由读者模仿完成。
*/
public class LoginActivity extends AppCompatActivity {
    static final int LOGIN_SUCCESS = 1;
    static final int USER_NOT_EXIST = -1;
    static final int PASSWORD_INCORRECT = -2;
    static String SERVICE_ROOT_URL = "http://192.168.11.191:8080/LoginServlet/"; //项目的
                                                                                   根路径
    static String SERVICE_UAP_URL = SERVICE_ROOT_URL + "images/UAP/"; //服务器用户头
                                                                       像路径
    static final String TAG = "测试";
    ImageView iv_UAP;
    EditText et_username;
    EditText et_password;
    Button btn_login;
    Button btn_register;
    SharedPreferences sharedPreferences;
    User user; //个人用户信息
    String username,password;
    int status；  //用户登录结果码
    ArrayList<User> friendsList; //好友用户信息集合
    String [] visitedList;  //在用户名文本框下显示使用共享存储的已登录过的用户名
    ListPopupWindow listPopupWindow;  //实现文本框的选择输入
    //Handler 对象，用于接收"登录子线程"发送过来的消息
    private Handler handler = new Handler(){
```

```java
        @Override
        public void handleMessage(Message msg) {
            String toastText = "";
            switch (msg.what){
                case LOGIN_SUCCESS:
                    toastText = "登录成功"; break;
                case USER_NOT_EXIST:
                    toastText = "用户名不存在"; break;
                case PASSWORD_INCORRECT:
                    toastText = "密码错误"; break;
                default:toastText = "未知错误（如 Web 服务器未启动或 URL 错误等）"; break;
            }
            Toast.makeText(LoginActivity.this, toastText, Toast.LENGTH_SHORT).show();
        }
    };

    @Override
    protected void onCreate(Bundle savedInstanceState) {
        super.onCreate(savedInstanceState);
        setContentView(R.layout.login_view);
        getSupportActionBar().hide();
        sharedPreferences = getSharedPreferences("login", Context.MODE_PRIVATE);
        String jsonString;
        if ((jsonString = sharedPreferences.getString("jsonString","empty")).equals("empty")){
            init(); //如果是第一次登录，执行登录界面操作
        }else {
            //自定义方法，将 JSON 字符串转为 JavaScript 对象后，给类成员赋值
            parseJsonString(jsonString);
            Intent intent = new Intent(LoginActivity.this,HomeActivity.class);
            //携带列表数据
            intent.putParcelableArrayListExtra("friendsList",friendsList);
            startActivity(intent);
            finish();    //销毁当前的 Activity
        }
    }
    private void init() {
        iv_UAP = findViewById(R.id.iv_UAP);
        et_username= findViewById(R.id.et_username);
        et_password= findViewById(R.id.et_password);
        btn_login = findViewById(R.id.btn_login);
        btn_register = findViewById(R.id.btn_register);
        btn_login.setOnClickListener(new View.OnClickListener() {   //登录
            @Override
            public void onClick(View v) {
                new Thread(new Runnable() {
                    @Override
```

```java
            public void run() {
                try {
                    //调用 login()方法进行登录操作
                    int result = login();
                    Message message = new Message();
                    switch (result){
                        case LOGIN_SUCCESS :
                            Log.d(TAG,"登录成功");
                            sharedPreferences.edit().putLong(user.getUsername(),
                                               user.getUserID()).commit();
                            message.what=LOGIN_SUCCESS;
                            handler.sendMessage(message);
                            Intent intent = new Intent(LoginActivity.this,
                                               HomeActivity.class);
                            //对象序列化后传输
                            intent.putParcelableArrayListExtra("friendsList",
                                               friendsList);
                            startActivity(intent);
                            finish();
                            break;
                        case USER_NOT_EXIST:
                            Log.d(TAG,"用户不存在");
                            message.what=USER_NOT_EXIST;
                            handler.sendMessage(message);
                            break;
                        case PASSWORD_INCORRECT:
                            Log.d(TAG,"密码错误");
                            message.what=PASSWORD_INCORRECT;
                            handler.sendMessage(message);
                            break;
                        default:
                            Log.d(TAG, "其他错误");
                            break;
                    }
                }catch (Exception e){
                    Log.d(TAG,"Exception:"+e.getMessage());
                }
            }
        }).start();
    }
});
btn_register.setOnClickListener(new View.OnClickListener() {    //注册
    @Override
    public void onClick(View v) {
        View view = View.inflate(LoginActivity.this,R.layout.register_view,null);
        Button btn_register_start = view.findViewById(R.id.btn_register_start);
```

```java
            final EditText et_username = view.findViewById(R.id.et_username);
            btn_register_start.setOnClickListener(new View.OnClickListener() {
                @Override
                public void onClick(View v) {
                    Toast.makeText(LoginActivity.this, "请完善注册功能",
                            Toast.LENGTH_SHORT).show();
                }
            });
            new AlertDialog.Builder(LoginActivity.this)
                    .setView(view)
                    .show();
            btn_register_start.setOnLongClickListener(new View.OnLongClickListener() {
                @Override
                public boolean onLongClick(View v) {      //隐藏功能：长按键后修改服务
                                                                              器 IP
                    String ipAddress = et_username.getText().toString();
                    Toast.makeText(LoginActivity.this, "原服务器地址\n"+SERVICE_
                                                                          ROOT_URL,
Toast.LENGTH_LONG).show();
                    LoginActivity.SERVICE_ROOT_URL =
                                            "http://"+ipAddress+":8080/LoginServlet/";
                    LoginActivity.SERVICE_UAP_URL = SERVICE_ROOT_URL +
                                                                         "images/UAP/";
                    HomeActivity.SERVICE_ROOT_URL =
                                            "http://"+ipAddress+":8080/LoginServlet/";
                    HomeActivity.SERVICE_UAP_URL = SERVICE_ROOT_URL +
                                                                         "images/UAP/";
                    Toast.makeText(LoginActivity.this, "新服务器地址\n"+SERVICE_
                                                                          ROOT_URL,
                                                          Toast.LENGTH_LONG).show();
                    return true;
                }
            });
        }
    });
    //两个编辑文本框的监听器，接口 TextWatcher 包含 3 个要实现的方法
    TextWatcher textWatcher = new TextWatcher() {
        @Override
        public void beforeTextChanged(CharSequence s, int start, int count, int after) { }
        @Override
        public void onTextChanged(CharSequence s, int start, int before, int count) { }
        @Override
        public void afterTextChanged(Editable s) {
            //输入了用户名密码后才允许登录
            if(!("".equals(et_username.getText().toString())) &&
                                !("".equals(et_password.getText().toString())) ){
```

```java
            btn_login.setBackgroundColor(getResources().getColor(R.color.colorBtn2));
            btn_login.setEnabled(true);
        } else {
            btn_login.setBackgroundColor(getResources().getColor(R.color.colorBtn1));
            btn_login.setEnabled(false);
        }
        //如果用户登录过，就在用户头像区域加载该用户头像
        long userID;
        Map<String, ?> map = sharedPreferences.getAll();
        if(map.containsKey(et_username.getText().toString())){
            userID = (Long) (map.get(et_username.getText().toString()));
            String imageUrl = SERVICE_UAP_URL +userID+".png";
            //使用 Android 图片加载框架 Glide 3.7.0
            Glide.with(LoginActivity.this)
                    .load(imageUrl)      //加载网络资源
                    .error(R.drawable.sun)    //任选
                    .into(iv_UAP);       //刷新图像控件
        }else{
            iv_UAP.setImageResource(R.drawable.sun); //加载本地资源
        }
    }
};
et_username.addTextChangedListener(textWatcher);
et_password.addTextChangedListener(textWatcher);
//用户名文本框辅助输入：为用户名输入框右边的那个三角图片设置了监听器
et_username.setOnTouchListener(new View.OnTouchListener() {
    @Override
    public boolean onTouch(View v, MotionEvent event) {
        // et.getCompoundDrawables()得到一个长度为 4 的数组，分别表示左右上下 4 张图片
        Drawable drawable = et_username.getCompoundDrawables()[2];
        //如果右边没有三角形图片，不再处理
        if (drawable == null)
            return false;
        //如果不是按下事件，将不再处理(onTouch 包含按下手指和抬起手指 2 个动作,
        //两个动作执行完毕就是一个 onClick)
        if (event.getAction() != MotionEvent.ACTION_UP)
            return false;
        //按下的区域是否在三角形图片区域内
        if (event.getX() > et_username.getWidth() - et_username.getPaddingRight() –
                              drawable.getIntrinsicWidth()){
            showListPopulWindow();
        }
        return false;
    }
});
```

```java
    }
    void showListPopulWindow(){
        Set<String> usernameSet = sharedPreferences.getAll().keySet();
        visitedList = new String[usernameSet.size()];
        int i = 0;
        for (String username:usernameSet) {
            visitedList[i++] = username;
        }
        listPopupWindow = new ListPopupWindow(this);
        listPopupWindow.setAdapter(new ArrayAdapter<String>(this,
                            R.layout.support_simple_spinner_dropdown_item,visitedList));
        listPopupWindow.setAnchorView(et_username);
        listPopupWindow.setModal(true);
        listPopupWindow.setOnItemClickListener(new AdapterView.OnItemClickListener() {
            @Override
            public void onItemClick(AdapterView<?> parent, View view, int position, long id) {
                et_username.setText(visitedList[position]);
                listPopupWindow.dismiss();
            }
        });
        listPopupWindow.show();
    }
    private int login(){   //用户登录，使用网络通信框架 Volley，返回登录状态码
        //登录信息
        username = et_username.getText().toString();
        password = et_password.getText().toString();
        //用户登录 URL
        String loginUrl = SERVICE_ROOT_URL + "Login";
        //成功得到服务器响应时的监听器
        Response.Listener listener = new Response.Listener<String>() {
            @Override
            public void onResponse(String s) {
                //将接送字符串保存起来，下次直接登录
                sharedPreferences.edit().putString("jsonString",s).commit();
                //解析从服务器获取到的 JSON 字符串，给类成员赋值
                parseJsonString(s);
            }
        };
        // 未成功得到服务器响应时的监听器
        Response.ErrorListener errorListener = new Response.ErrorListener() {
            @Override
            public void onErrorResponse(VolleyError volleyError) {
                Log.d(TAG, "ErrorListener:"+volleyError.getMessage());
                Toast.makeText(LoginActivity.this, "服务器正在维护中...",
                                            Toast.LENGTH_SHORT).show();
            }
```

```java
        };
        //创建 StringRequest 对象
        StringRequest stringRequest= new StringRequest(Request.Method.POST,
                                            loginUrl,listener,errorListener){
            @Override
            protected Map<String, String> getParams() { //重写方法，POST 方式传参数
                Map<String,String> map = new HashMap<>();
                map.put("username",username);
                map.put("password",password);
                return map;
            }
        };
        RequestQueue requestQueue = Volley.newRequestQueue(this);
        //加入请求至队列
        requestQueue.add(stringRequest);
        try {
            Thread.sleep(1000);   // 休眠 1s，等待 status 值更新
        } catch (InterruptedException e) {
            e.printStackTrace();
        }
        return status;
    }
    //解析从服务器获取到的 JSON 字符串，给类成员赋值
    private void parseJsonString(String s) {
        try {
            //将 JSON 字符串转化为 JSON 对象
            JSONObject resultObject = new JSONObject(s);
            //设置用户
            JSONObject userObject = resultObject.getJSONObject("user");
            user = new User(userObject.getInt("userID"), userObject.getString("username"));
            //设置该用户的好友
            JSONArray friendsArray = resultObject.getJSONArray("friends");
            friendsList = new ArrayList<User>();
            if (friendsArray.length() > 0) {
                for (int i = 0; i < friendsArray.length(); i++) {
                    JSONObject friendObject = friendsArray.getJSONObject(i);
                    long userID = friendObject.getLong("userID");
                    String username = friendObject.getString("username");
                    User user = new User(userID, username);
                    friendsList.add(user);
                }
            }
            //设置用户的登录状态
            status = resultObject.getInt("status");
        } catch (JSONException e) {
            Log.d(TAG, "JSONException:" + e.getMessage());
```

            }
        }
}
```

注意：JSON 字符串与 JSON 对象可以相互转化。

### 11.3.3　App 的主界面程序设计

登录成功后进入好友界面，其 HomeAtivity 代码如下：

```java
public class HomeActivity extends AppCompatActivity {
    //下面的地址，要与 LoginActivity 里一致！
    static String SERVICE_ROOT_URL = "http://192.168.11.191:8080/LoginServlet/";
    static String SERVICE_UAP_URL = SERVICE_ROOT_URL + "images/UAP/"; //服务器用户头
                                                                      像路径

    ArrayList<User> friendsList;
    @Override
    protected void onCreate(@Nullable Bundle savedInstanceState) {
        super.onCreate(savedInstanceState);
        setContentView(R.layout.home_view);
        getSupportActionBar().hide();
        friendsList = getIntent().getParcelableArrayListExtra("friendsList");
        myBaseAdapter adapter = new myBaseAdapter();
        ListView lv_friends = findViewById(R.id.lv_friends);
        lv_friends.setAdapter(adapter);
    }
    class myBaseAdapter extends BaseAdapter{
        @Override
        public int getCount() {
            return friendsList.size();
        }
        @Override
        public Object getItem(int position) {
            return friendsList.get(position);
        }
        @Override
        public long getItemId(int position) {
            return position;
        }
        @Override
        public View getView(int position, View convertView, ViewGroup parent) {
            View view = View.inflate(getApplicationContext(),R.layout.home_view_listitem, null);
            TextView tv_username = view.findViewById(R.id.tv_username);
            ImageView iv_ico = view.findViewById(R.id.iv_ico);
            User user = (User) getItem(position);
            tv_username.setText(user.getUsername());
            iv_ico.setImageResource(R.drawable.sun);
            String imageUrl = SERVICE_UAP_URL +user.getUserID()+".png";
```
```

```java
                    Log.d("测试", "imageUrl:"+imageUrl);
                    Glide.with(HomeActivity.this)
                            .load(imageUrl)
                            .error(R.drawable.sun)    //任选
                            .into(iv_ico);
                    return view;
                }
            }
    @Override
    public void onBackPressed() {    //按返回键时
        AlertDialog.Builder builder = new AlertDialog.Builder(this);
        builder.setTitle("确认退出吗？")
                .setNegativeButton("退出", new DialogInterface.OnClickListener() {
                    @Override
                    public void onClick(DialogInterface dialogInterface, int i) {
                        finish();
                    }
                })
                .setPositiveButton("注销", new DialogInterface.OnClickListener() {
                    @Override
                    public void onClick(DialogInterface dialogInterface, int i) {
                        //删除 jsonString 记录
                        getSharedPreferences("login", Context.MODE_PRIVATE).edit()
                                                    .remove("jsonString").commit();
                        Intent inttent = new Intent(HomeActivity.this, LoginActivity.class);
                        startActivity(inttent);
                        finish();
                    }
                })
                .create()
                .show();
    }
}
```

按返回键时，出现如图 11.3.5 所示的对话框。

在退出对话框中，退出与注销的功能差别如下：

● 退出不会清除登录记录，只是退出应用，下次进入应用时会直接到好友界面；

● 注销会清除登录记录，并返回登录界面。

图 11.3.5　用户退出和注销对话框

习 题 11

一、判断题

1．HttpURLConnection 编程是基于 HTTP 请求的网络编程。
2．Web 服务器程序返回的数据格式只能是 JSON。
3．Google Gson 包用于序列化 Java 对象为 JSON 字符串。
4．在 Android 中，使用 Handler 对象发送消息，只能在子线程中进行。
5．在 Activity 程序里使用 Glide 框架加载网络图像，会阻塞 UI 线程。

二、选择题

1．手机 App 与 Web 服务器通信时，Web 服务器的类型可以是____。
　　A．JSP　　　　　B．PHP　　　　　C．ASP.NET　　　　　D．均可
2．HttpURLConnection 与 URL 位于相同的软件包____里。
　　A．java.os　　　B．java.net　　　C．java.sql　　　　　D．java.util
3．Google Gson 包的类 Gson 提供了转 Java 对象为 JSON 字符串的方法____。
　　A．toString()　　B．wait()　　　C．equals()　　　　　D．toJson()
4．使用网络框架 Volley 时，必须创建的对象是____。
　　A．StringRequest　　　　　　　　B．ImageRequest
　　C．RequestQueue　　　　　　　　D．JsonObjectRequest
5．在 Android Studio 项目里，加载 Volley 框架，应在 Project Struture 对话框里，先选择模块，后选择____选项。
　　A．Dependencies　　　　　　　　B．Properties
　　C．Flavors　　　　　　　　　　　D．Signing

三、填空题

1．使用 URL 对象的 openConnection()方法，可以得到一个____对象。
2．使用 HttpClient 对象的____方法，可以得到一个 HttpResponse 对象。
3．基于 Http 的 Android 通信项目，必须在清单里注册____权限。
4．在 Glide 框架里，静态方法 Glide.with()的返回值类型为____。
5．使用 Glide 框架的 RequestManager 的 load()加载图像、____方法刷新图像控件。

实 验 11

一、实验目的

（1）掌握使用 HttpURLConnection 获取网络信息资源的方法。
（2）掌握使用接口 HttpClient 调用 Web 服务的方法。
（3）掌握 Android 网络框架 Glide 和 Volley 的使用。
（4）掌握手机客户端与 Web 服务器通信的编程方法。

二、实验内容及步骤

【预备】在 Android Studio 中，新建名为 Example11 的项目后，访问本课程配套的网站 http://www.wustwzx.com/as/sy/sy11.html，复制相关代码，完成如下几个模块的设计。

1. 使用 HttpURLConnection 访问网络资源（抓取网页）

（1）在项目 Example 11 里，新建名为 example11_1 的模块，在清单文件里，注册访问 Internet 网络的权限 android.permission.INTERNET。

（2）在布局文件里，分别定义一个 Button 类型和一个 TextView 类型的控件作为类的成员；在 MainActivity 里，定义 AsyncTask 的子类作为内部类，用以获取网络数据。

（3）在 MainActivity.java 的 onCreate()方法中，实例化 2 个成员对象，编写按钮的单击事件事件监听器，调用异步任务类获取网络数据。

（4）在异步任务类中重写方法 doInBackground()，使用 Java 字符流操作获取网络数据；在方法 onPostExecute()中刷新 UI 控件。

（5）部署本模块，做运行测试。

2. 使用接口 HttpClient 调用 Web 服务（手机归属地查询，参见例 11.1.2）

（1）打开浏览器，在地址栏里输入访问 Web 服务的网址：http://ws.webxml.com.cn/WebServices/MobileCodeWS.asmx/getMobileCodeInfo?mobileCode=1597205&userID=，查看返回的 XML 数据包含了手机归属地信息。

（2）在项目 Example 11 里，新建名为 example11_1a 的模块，在清单文件里，注册访问 Internet 网络的权限 android.permission.INTERNET。

（3）对本模块，添加 HttpClient 的依赖包。

（4）在布局文件里，分别定义 1 个 EditText 类型和 1 个 Button 类型的控件作为类的成员。

（5）在 MainActivity.java 的 onCreate()方法里，强制使用子线程策略；定义按钮的单击事件监听器，获取网络的 XML 数据并解析。最后，刷新显示结果信息的 TextView 控件。

（6）部署本模块，做运行测试。

第 11 章　Android 网络编程

（7）在浏览器地址栏里，输入：

https://chongzhi.jd.com/json/order/search_searchPhone.action?mobile=15972052815，查看返回手机归属地的 JSON 数据。

（8）在项目里，新建名为 example11_1b 的模块，与抓取网页模块类似，使用异步任务类获取网络数据，最后做运行测试（请读者自行完成）。

3．使用 Volley 框架获取豆瓣网站里的电影信息（参见例 11.2.2）

（1）访问豆瓣网 https://api.douban.com/v2/movie/subject/26842702，返回 JSON 字符串，通过 JSON 在线格式化工具，可查看到表示电影的文本和图像的 JSON 信息。

（2）在项目 Example 11 里，新建名为 example11_2 的模块，在清单文件里，注册访问 Internet 网络的权限 android.permission.INTERNET。

（3）在布局文件里，分别定义用于输入电影 ID 的 EditText 控件和用于显示电影信息的 TextView 控件和 ImageView 控件等作为类的成员。

（4）对本模块添加 Volley 框架的依赖包。

（5）在 MainActivity.java 里，定义 AsyncTask 的子类作为内部类。在其重写的方法 doInBackground() 里，使用 Volley 框架创建获取网络资源的 JsonObjectRequest 对象，在其回调方法里创建下载网络图像资源的 ImageRequest 对象。

（6）在 MainActivity.java 的 onCreate() 方法里，实例化类的成员对象，编写按钮的单击事件事件监听器，执行异步任务获取网络数据。最后，刷新显示结果信息的相关控件。

（7）部署本模块，做运行测试。

4．手机客户端程序与 Web 服务器的信息交互（参见第 11.3 节）

（1）在 Eclipse 中，创建名为 LoginServlet 的 Java Web 项目。

（2）创建名为 Login 的 Servlet 程序，使用 Google Gson 包序列化 Java 对象为 JSON 字符串，返回表示用户登录结果的 JSON 字符串。

（3）确保手机 IP 与计算机 IP 在同一网段内，并记录好计算机的 IP。

（4）在项目 Example 11 里，新建名为 example11_3 的模块。

（5）创建实体类 User，并实现序列化接口 Parcelable。

（6）对本模块同时引入 Volley 和 Glide 2 种框架。

（7）编写手机登录程序，访问 Web 服务器项目 LoginServlet，实现用户登录功能。

（8）编写主 Activity，展示用户登录成功后的好友列表。

（9）部署本模块，做运行测试。

（10）模仿用户登录的实例，完成用户注册功能。

三、实验小结及思考

（由学生填写，重点写上机中遇到的问题。）

习题答案

习 题 1

一、判断题(正确用"T"表示,错误用"F"表示)

1~6：F T F T T T

二、选择题

1~5：B B B C D

三、填空题

1. 四 2. Java 3. Ctrl+Alt+O 4. 5554 5. View→Tool Windows

习 题 2

一、判断题(正确用"T"表示,错误用"F"表示)

1~5：T F F T F

二、选择题

1~6：C D A D C D

三、填空题

1. package 2. @ 3. label 4. 主 5. setContentView()

习 题 3

一、判断题(正确用"T"表示,错误用"F"表示)

1~5：T T F T F 6~9：F F T T

二、选择题

1~5：A D C D C

三、填空题

1．onCreate() 2．setImageResource() 3．sp 4．onDestrory() 5．ListAdapter

习 题 4

一、判断题(正确用"T"表示，错误用"F"表示)

1~5：F F T T F

二、选择题

1~5：B A B C C

三、填空题

1．getBundleExtra() 2．READ_EXTERNAL_STORAGE 3．结果
4．startActivityForResult() 5．Environment

习 题 5

一、判断题(正确用"T"表示，错误用"F"表示)

1~7：T T F T T F F

二、选择题

1~5：B B A C D

三、填空题

1．android.app 2．隐式 3．IBinder 4．AIDL 5．onBind()

习 题 6

一、判断题(正确用"T"表示，错误用"F"表示)

1~7：F T F T T T T

二、选择题

1~5：C D A A B

三、填空题

1．<receiver>　2．onReceive()　3．<action>　4．BroadcastReceiver　5．Intent

习　题　7

一、判断题(正确用"T"表示，错误用"F"表示)

1～5：T　F　F　T　T

二、选择题

1～6：C　D　A　C　D　B

三、填空题

1．db　2．SQLiteOpenHelper　3．SQLiteDatabase　4．void　5．onUpgrade()
6．boolean

习　题　8

一、判断题(正确用"T"表示，错误用"F"表示)

1～6：T　F　T　F　T　T

二、选择题

1～6：A　D　C　C　A　C

三、填空题

1．<provider>　2．逻辑　3．content　4．ContentResolver　5.doInBackground()

习　题　9

一、判断题(正确用"T"表示，错误用"F"表示)

1～7：F　F　T　T　F　T　T

二、选择题

1～5：C　C　D　A　D

三、填空题

1．4　2．startScan()　3．getConnectionInfo()　4．getDefaultAdapter()　5．UUID

习 题 10

一、判断题(正确用"T"表示,错误用"F"表示)

1~5: F F T T T

二、选择题

1~5: C A C B A

三、填空题

1. requestLocationUpdates()   2. GPS
3. ACCESS_COARSE_LOCATION   4. MapView

习 题 11

一、判断题(正确用"T"表示,错误用"F"表示)

1~5: T F T T F

二、选择题

1~5: D B D C A

三、填空题

1. URLConnection 或 HttpURLConnection   2. execute()   3. INTERNET
4. RequestManager   5. into()

参 考 文 献

[1] 张晓龙，吴志祥，刘俊. Java 程序设计简明教程[M]. 北京：电子工业出版社. 2018.
[2] 吴志祥，柯鹏，张智，胡威. Android 应用开发案例教程[M]. 武汉：华中科技大学出版社. 2015.
[3] 吴志祥，张智，曹大有，焦家林，赵小丽. Java EE 应用开发教程[M]. 武汉：华中科技大学出版社. 2016.
[4] 李波，史江萍，王祥凤. Android 4.X 从入门到精通[M]. 北京：清华大学出版社. 2012.
[5] 王向辉，张国印，赖明珠. Android 应用程序开发（第 2 版）[M]. 北京：清华大学出版社. 2012.